T0185616

SpringerBriefs in Molecular Science

Chemistry of Foods

Series Editor
Salvatore Parisi, Al-Balqa Applied University, Al-Salt, Jordan

The series Springer Briefs in Molecular Science: Chemistry of Foods presents compact topical volumes in the area of food chemistry. The series has a clear focus on the chemistry and chemical aspects of foods, topics such as the physics or biology of foods are not part of its scope. The Briefs volumes in the series aim at presenting chemical background information or an introduction and clear-cut overview on the chemistry related to specific topics in this area. Typical topics thus include:

- Compound classes in foods—their chemistry and properties with respect to the foods (e.g. sugars, proteins, fats, minerals, …)
- Contaminants and additives in foods—their chemistry and chemical transformations
- Chemical analysis and monitoring of foods
- Chemical transformations in foods, evolution and alterations of chemicals in foods, interactions between food and its packaging materials, chemical aspects of the food production processes
- Chemistry and the food industry—from safety protocols to modern food production

The treated subjects will particularly appeal to professionals and researchers concerned with food chemistry. Many volume topics address professionals and current problems in the food industry, but will also be interesting for readers generally concerned with the chemistry of foods. With the unique format and character of SpringerBriefs (50 to 125 pages), the volumes are compact and easily digestible. Briefs allow authors to present their ideas and readers to absorb them with minimal time investment. Briefs will be published as part of Springer's eBook collection, with millions of users worldwide. In addition, Briefs will be available for individual print and electronic purchase. Briefs are characterized by fast, global electronic dissemination, standard publishing contracts, easy-to-use manuscript preparation and formatting guidelines, and expedited production schedules.

Both solicited and unsolicited manuscripts focusing on food chemistry are considered for publication in this series. Submitted manuscripts will be reviewed and decided by the series editor, Prof. Dr. Salvatore Parisi.

To submit a proposal or request further information, please contact Tanja Weyandt, Publishing Editor, via tanja.weyandt@springer.com or Prof. Dr. Salvatore Parisi, Book Series Editor, via drparisi@inwind.it or drsalparisi5@gmail.com

More information about this subseries at http://www.springer.com/series/11853

Rachid Chaib · Michele Barone

Chemicals in the Food Industry

Toxicological Concerns and Safe Use

 Springer

Rachid Chaib
Department of Transport Engineering
University of Constantine 1
Constantine, Algeria

Michele Barone
Associazione "Componiamo il Futuro"
(CO.I.F.) Palermo
Palermo, Italy

ISSN 2191-5407 ISSN 2191-5415 (electronic)
SpringerBriefs in Molecular Science
ISSN 2199-689X ISSN 2199-7209 (electronic)
Chemistry of Foods
ISBN 978-3-030-42942-3 ISBN 978-3-030-42943-0 (eBook)
https://doi.org/10.1007/978-3-030-42943-0

This Springer imprint is published by the registered company Springer Nature Switzerland AG
The registered company address is: Gewerbestrasse 11, 6330 Cham, Switzerland

To my Mother

You have always been able to transmit to me all the moral and human values that you believe in: justice, perseverance, rigor, self-giving, love. You have always passed our happiness before yours; your courage in difficult times is exemplary.

To my Wife

I draw on your trust and your love my daily motivation; thank you for allowing me to build this wonderful path by your side and all the proofs of affection that you carry me daily.

In memory of my Father

Prof. Rachid Chaib

Preface

Chemicals are essential for the production of a large number of industrial goods. Many industrial sectors use chemicals (cleaning, degreasing, paints, inks, adhesives, foods, photocopiers, cleaning products, etc.). These substances gradually invade the environment and many aspects of our life. It is through these products that millions of people can produce and consume goods necessary for life. Unfortunately, raising concerns about their effects on health and the environment are emerged. Poorly used, without basic or special precautions, many of these chemicals can become dangerous and sometimes cause great accidents and serious diseases. More than 33 % of employees in the general industry, including food and beverage articles, are exposed to at least one chemical product during their work. Rejected into nature without precautions or treatments, chemicals pollute the environment and break the natural balance for the development of life on earth, in water and air.

Some of these substances can have significant adverse effects on the environment and health, even at low doses. Other products also raise concerns because of their persistence in environments, difficulty of measuring them, and the lack of knowledge of their environmental and health impacts on a long-term scale. They can also be emitted as dust, gas, vapors, and their presence can then be unsuspected resulting in more serious injuries and/or material damages. It should also be noted that occupational cancers cause a mortality rate higher than that caused by occupational accidents: The multiplication of used chemicals and the complexity of industrial processes therefore really require increased vigilance regarding the carcinogenic risk they induce. Food industries are not an exception. Cancer is now the leading cause of death in industrialized countries. Respiratory cancers are the most frequent occupational cancers: Cancers attributable to asbestos, benzene, ionising radiations, and wood dust alone account for nearly 90% of work-related cancers. However, these diseases often take several decades to develop after exposure; their link with work is therefore not always easy to establish. Employers and workers must be able to protect themselves from carcinogens. In this respect, chemicals play a central role in health and safety at work. Therefore, the consideration of health issues at work can be an opportunity to change the medium-term performance of all areas of industries; on the other hand, failure to take these issues into account may penalize companies in future years.

Since 1930s, the global chemical production has increased by a factor of 400. Of the 100,000 chemical substances in the European Union, 30,000 are produced at more than one ton per year. However, less than 3000 products have been thoroughly analyzed for their dangerousness when speaking of toxicity and ecotoxicity. One of the main reasons has been the entry of chemistry into food and beverage production activities.

With the increasing complexity of the industrial enterprise, the constant presence of chemicals, and the rapid evolution of little and medium companies into large enterprises, risk assessment becomes a critical and strategic response when speaking of health and safety. The same thing can be surely affirmed when speaking of industries producing or manipulating foods, beverages, and related intermediates, because chemicals can be:

(a) Additives and/or
(b) Components of food-contact materials and objects
(c) Agents for non-food production such as sanitizers and cleaning substances (and consequently unavoidable in a safe food production environment), and/or
(d) Compounds used for analytical determinations into laboratories annexed to food plants.

On the one hand, qualified workforce has to be maintained; on the other hand, 'no one should lose his life trying to win it'. Now, awareness of health and safety issues at work is increasing throughout society as work-related accidents and occupational diseases have a huge impact on workers' health and economic and social benefits. Companies have gradually come to consider these concerns within their organization, and this consciousness has become evident in food environments. It is therefore essential that young managers, future employees, and managers master safety regulations: Safety is a priority of industrial activities. It is a positive cultural element that allows other improvements in the business. A management which cannot manage the safety and health of its workers is not able to manage other functions.

In our experience, health and safety at work does not mean much for some employers, employees, and their representatives. Moreover, with the evolution of the work, even its risks, it becomes more and more insufficient to establish general safety rules, relying solely on compliance with the posters of standards and regulations. It is mandatory to gradually bring companies to consider these concerns within their organization. As a result, susceptibility to health and safety issues at work is increasing throughout the society.

This book is intended for 'Licence–Master–Doctorate' (LMD) students from all sectors, especially those with a predominantly industrial background and, more particularly, students specializing in food hygiene, public safety, and environment. In fact, future graduates and masters will have in their practice within the company to have incorporated good notions in various topics. In this ambit, the food and beverage production is specially considered.

The danger of a chemical product is an intrinsic property of compounds. No direct action can therefore modify this property without changing the nature of the product. Exposure to the product is the set of handling conditions for this product that may cause a target's exposure to the adverse effects of that product. By an action on the modalities of this exposure, it is possible to avoid or mitigate it and thus to act on the probability of an incident or an accident.

As a result, it has become essential to give all future executives a true 'security' spirit that will enable them to plan and act very effectively within our companies.

The role of the professional in industrial hygiene and safety is the recognition and evaluation of the industrial environment when it is compared with the need to control a dangerous exposure. The work of industrial hygiene and safety professionals involves four main functions. These functions are:

(1) Recognition of potential hazards in the work environment
(2) Measurement of the work environment with the aim of estimating the presence of a hazard and the subsequent evaluation of measures taken to determine whether a hazard exists
(3) Identification and recommendation of controls that can be implemented
(4) Elimination or reduction of exposure to hazards for the worker
(5) Anticipation/prevention of potential hazards, if possible.

The number and variety of chemicals present in different industries and in laboratories continue to increase. Dangers of these products are not always well known, even by specialists. Irritant, toxic, flammable, carcinogenic, reprotoxic, etc., they must be handled with precaution, so that staff, researchers, doctors, technicians, teachers, and students can work safely for their activity and their health in the best possible conditions of hygiene and safety. Also, chemicals are part of our daily lives. Unfortunately, the use and manufacturing of these synthetic chemicals are causing a growing number of health problems in humans. Finally, let us not forget that the field of health and safety at work is continuously evolving.

Unfortunately, users of chemical products often underestimate their ignorance of the dangerousness of these products and related exposure risks. Health consequences are highly variable, ranging from temporary disability to death to incapacity to work or disability. They can be sudden (burns, asphyxiation), brutal (acute intoxication, fire, explosion), or take shape of illness or chronic intoxication. They can occur gradually or appear several years after exposure.

Health and safety at work must be a value we share.

Constantine, Algeria   Prof. Rachid Chaib
Palermo, Italy   Mr. Michele Barone

Contents

Chapter 1
Chemicals in the Food and Beverage Industry: An Introduction

Abstract All products present in a company can be a source of danger for workers and the environment, and this statement is particularly true for incoming products, manufactured products, waste, and chemicals. A large number of chemical products can become potentially dangerous and be sometimes at the origin of notable accidents or serious diseases. All companies, from the largest to the most modest enterprise, must take this aspect into consideration, including the sector of foods and beverages. Seven million people worldwide die each year because of certain diseases that are favoured by indoor and outdoor air pollution. This chapter is a general introduction to chemicals in the industry (food and non-food sectors) in relation to their importance, chemical risks, occupational exposure limits, pollution events, and toxicological evaluations.

Keywords Chemical risk · Endocrine disruptor · European Union · Occupational exposure limit · Pollution · Threshold limit value · Toxicology

Abbreviations

ATA	American Tinnitus Association
AISS	Association Internationale de la Sécurité Sociale
AEV	Average exposure value
CO_2	Carbon dioxide
CMR	Carcinogenic, mutagenic, reprotoxic
CCHST	Centre canadien d'hygiène et de sécurité au travail
CNRS	Centre National de la Recherche Scientifique
CAS	Chemical abstracts service
DCA	Dangerous chemical agent
EU	European Union
SSTI	Fédération régionale des services de santé au travail interentreprises
FAO	Food and Agriculture Organization of the United Nations
DGUV	German Social Accident Insurance

© The Author(s), under exclusive license to Springer Nature Switzerland AG 2020
R. Chaib and M. Barone, *Chemicals in the Food Industry*,
Chemistry of Foods, https://doi.org/10.1007/978-3-030-42943-0_1

HF	Hydrofluoric acid
INRS	Institut National de Recherche et de Securitè
IARC	International agency for research on cancer
LD_{50}	Lethal dose 50
LC_{50}	Mean lethal concentration
OEL	Occupational exposure limit
STEL	Short-term exposure limit
Threshold limit value	TLV
UNEP	United Nations Environment Programme
WHO	World Health Organization

1.1 Introduction to Chemicals

All products present in a company can be a source of danger for employees and the environment, and this statement is particularly true for incoming products, manufactured products, waste, and chemicals. People work, produce, and consume goods necessary for life thanks to chemicals. On the other hand, when badly used (without elementary or special precautions), a large number of chemical products can become dangerous and be sometimes at the origin of notable accidents or serious diseases. They pollute the environment and break the natural balance necessary for the development of life on earth, in water, and in air. Chemicals are substance of general use nowadays in all sectors of anthropic activities, including industrial enterprises (cleaning, degreasing, paints, inks, adhesives, foods, etc.) as well as outside industry. It is therefore all workers who must feel concerned by this problem and be attentive to their chemical risks. These substances can also be emitted in the form of dust, gases, vapors, and their presence can then be surprising with possible body injuries and/or material damages.

All companies, from the largest to the most modest enterprise, must take this aspect into consideration. Even if they look familiar, used without care, they can cause accidents or trigger occupational diseases. According to the latest World Health Organization (WHO) modelizations, seven million people worldwide die each year because of certain diseases (cardiovascular diseases, stroke, lung cancer, respiratory infections, etc.) that are favoured by indoor and outdoor air pollution. Many products are composed of substances that can be dangerous (WHO 2016).

Whatever the route of penetration is, chemical products can pass into the blood and then into the body. It should be noted that disorders do not always appear immediately. As a result, the chemical risk presents features justifying a different approach and search for information if compared to other risks. In particular, occupational accidents and diseases should not be considered as one burden among others, but as a 'malfunction.' The danger does not come solely from the product, but from a lack of knowledge of risks associated with poor use conditions. Recent studies have shown that more than 200 substances in industrial settings are potentially ototoxic (ATA 2016). When a chemical has an effect on the auditive system, it is said to be ototoxic. Among these substances are solvents, the main ones being toluene, styrene, xylene, carbon disulfide, and trichlorethylene; asphyxiant compounds including carbon monoxide and hydrogen cyanide; metals including lead and mercury; and pesticides such as organophosphates.

More than 159 million of organic and inorganic substances are present in the scientific literature, according to Chemical Abstracts Service (CAS) and with an increasing exponential rate of growth (Binetti et al. 2008; CAS 2020). More than 105 chemicals are registered and marketed in Europe, 10% of which are sold in quantities of more than ten tonnes per year and 30% in quantities exceeding one metric ton. It should be noted that the general term of product or compound covers two sets: substances and preparations, terms referring, respectively, to 'pure' products and mixtures of two or more substances.

1.2 Chemicals in General

Hazardous products are chemicals having dangerous properties potentially able to cause injury, damage, or harm to people, facilities, or the environment. The use of these products, which are ever more numerous in all industrial, artisanal, and agricultural sectors, exposes the most part of workers to risks of acute or chronic toxicity, by respiratory, cutaneous, or digestive means. Many of these products—solids, dust, liquids, gases, vapors, or fumes—are corrosive, irritant, allergenic, asphyxiating, fibrogenic, carcinogenic, reprotoxic, etc., and sometimes in low doses and reduced exposure times. Used alone or in mixtures, chemicals may present various effects harmful to human health. A workplace can bring together different people and activities (factories, workplaces, offices, hospitals, farms, etc.).

In an industrial process, production takes place in different stages and operations, as the raw materials are transformed into finished products with the possible use of chemicals. Some of these products that we use at work can enter our body and cause damage. Damage to health is not always visible immediately and sometimes only appears years later. This is the case, for example, with carcinogenic, mutagenic, or reprotoxic substances (toxic for reproduction).

The diversity of occupational exposures and other factors involved in the development of cancers (food, tobacco, environment, genetic predisposition, etc.) is such that it is often very difficult to prove the link between cancer and cause (professional activity), especially since the disease often occurs years after exposure or even after the cessation of professional activity. In many cases, linking cancer to professional causes requires a careful retrieval of the history of exposures experienced by the single worker. Studies have identified a number of occupational factors potentially able to increase risks or promote cancer development (Kasbi-Benassouli et al. 2005). They include asbestos, arsenic, benzene, chromium, vinyl chloride, nickel, aromatic amines, polycyclic aromatic hydrocarbons, wood or leather dust, and ionizing radiations. Because their potential to cause cancer or increase its frequency is proven, these substances are part of 'carcinogenic, mutagenic, reprotoxic' (CMR) agents (Havet et al. 2017). The removal or substitution of these products is required, whenever it is technically possible. Among the main types of cancer which a link has been established for with substances in the occupational environment, we can mention:

(1) Lung cancer (asbestos and other cancerogen products)
(2) Mesothelioma (the main cause: asbestos)
(3) Cancer of the nasal cavity (wood dust, nickel, chromium, and arsenic)
(4) Bladder cancer (aromatic amines, coal tars, and arsenic)
(5) Leukemias (benzene, ionizing radiations, and certain pesticides).

Dangerous chemical agents (DCA) are compounds with recognized hazardous characteristics (Health and Safety Authority 2020). Some of them may be explosive, flammable, toxic, and/or corrosive. They may also be chemical agents posing a risk to the health and safety of workers because of their physicochemical, chemical, or toxicological properties. These hazardous chemicals include CMR: Consequently, they can cause cancer, induce mutagenic effects on the body of the exposed person or on the fetus of a pregnant woman, or cause fertility or sterility problems (Kasbi-Benassouli et al. 2005).

In view of the dangers they present, these classified substances and mixtures are subjected to restrictive regulations, in particular on the work field, and their use in food and beverage environments is not a pure exception. It is essential to identify them, that is to say, to make an inventory of products used and working situations that can give rise to such exposures, as shown in Fig. 1.1. The use of similar substances when speaking of edible products for human and animal nutrition has to be considered carefully. At first sight, the impression of food consumers would be that edible product has not prepared, processed, stored, manipulated, or transported with the concomitant presence of hazardous components such as explosive substances. The truth is completely different.

Thus, detailed information on relevant processes, operations, and other activities is needed with the aim of identifying agents used, including raw materials, materials handled or added during manufacturing, primary products, intermediate products, finished, reactional, and by-products. It may also be interesting to identify additives and catalysts involved in a process. Raw materials and filler materials that are solely known by their trade name should be evaluated based on their chemical composition and the safety or information sheets normally available from the manufacturer or supplier.

As a result, the term use of chemicals at work applies to any occupation that could expose a worker to a chemical, including

(1) Production
(2) Handling
(3) Storage
(4) Transport

Chemicals in the Industry today...

Chemical Substances (pure form)	Chemical mixtures or preparations (more than one single substance)
Examples: acetone, ethanol, ...	Examples: coatings, adhesives, polymers, ...

Fig. 1.1 Chemical substances and mixtures are subjected to restrictive regulations, in particular on the work field. It is essential to consider differences between single (pure) substances and preparations or mixtures. This table has been realized by Carmelo Parisi, currently a student at the Liceo Scientifico Stanislao Cannizzaro, Palermo, Italy

(5) Disposal and treatment of chemical waste
(6) Emission of chemicals resulting from professional activities
(7) Maintenance, repair, and cleaning of equipment and containers used for chemicals.

1.3   Several Definitions

1.3.1   Definition 1

Chemical risk refers to all hazardous situations involving chemicals, in all possible situations (no exclusions!). It groups together the risks associated with storing products, transporting them in and out of the food and non-food company, handling them, and also eliminating them (United States Department of Labor 2013).

Therefore, the chemical risk is the danger that a product represents by its chemical properties, by its conditions of use or by its limit value of occupational exposure.

The chemicals we use at home or in our work often seem familiar and innocuous, especially in food and beverage environments. Attention and vigilance are needed!

Chemicals can be in three states: solid, liquid, or gaseous. They are often used deliberately for everyday use but can also be emitted in the form of dust, gases, and vapors, and their presence may be undesirable. These products are classified according to the risks they present (Parisi 2016):

(1) Physicochemical properties: flammable, unstable products that may give rise to fires, explosions, or reactive actions (heat release, gas emission, product projection)
(2) Toxicological properties: harmful, corrosive, toxic, irritant, carcinogenic, etc., having adverse effects on human health (intoxication) and on the environment (pollution).

These substances can cause:

(1) On the one hand, serious accidents (burns, intoxications, explosions, etc.) or even the origin of occupational diseases (allergies, cancers, etc.)
(2) On the other hand, environmental pollution during accidental spills or diffusion, and/or chronic releases toward the natural environment, with the effect of jeopardizing the equilibrium of the ecosystem.

Mandatory regulations have established classification and labeling rules based on these properties (hazard class). Any chemical that meets these rules is called DCA.

1.3.2 Definition 2

A chemical agent is a chemical product, generally marketed and often subjected to labeling, by addition of chemical elements and/or compounds and related mixtures, without relation to natural or artificial states. This substance may be supplied, stored, and used in the form of vapors, dust, smoke, etc. In addition, chemical sub-products of each industrial or artisanal process include chemical waste, without relation to the intentional or unintentional production and the final destination (United States Department of Labor 2013).

A DCA is essentially (European Commission 2013):

(1) Any agent which is subjected to a specific regulatory marking (explosive, oxidizing, highly toxic, toxic, harmful, corrosive, irritant, sensitizing, carcinogenic, mutagenic, and toxic to reproduction agents)
(2) Any chemical agent which could be recognized to have occupational exposure limit values (AISS 2016)
(3) Any chemical agent which may present a risk for the health and safety of workers because of its physicochemical, chemical, or toxicological properties, without relation to above-mentioned classification criteria and its homogeneous or heterogeneous state. Examples in this ambit: formwork oil, epoxy resins, tars, iron oxide, etc.

These definitions finally make possible to introduce into any inventory of dangerous products such as dust produced by certain mechanical operations (wood dust) which were known to be carcinogenic, without being classifiable. We can also cite fumes from thermal operations such as welding, melting, and baking of certain materials including those containing lead or chromium. In the ambit of food and beverage industries, the problem is extremely important because many (or all) of food preparation processes include mechanical operations, low- or high-temperature steps, melting/fusion, baking, freezing operations, smoking operations (where wood dust is generated or liquefied by means of water impregnation).

The danger of a chemical agent is a property of the latter which specifies the type of damage it may cause: intoxication, irritation, injury, burn, cancer, fire, explosion, etc. It is indicated in the labeling when it exists.

1.3.3 Definition 3

'Substances' mean chemical elements and their compounds as they occur naturally or as they are obtained by any process of production possibly containing any additive necessary to preserve the stability of the product and any impurity resulting in the process, excluding any solvent that can be separated without affecting the stability of the substance or altering its composition (United States Department of Labor 2013).

1.3.4 Definition 4

'Preparations' mean mixtures or solutions consisting of two or more substances, instead of one (and homogeneous) element or substance (Csuros 1997). An important example is the so-called food gas: foaming agents, packaging gases, propellants, and raising agents (Laganà et al. 2019a, b, c, d).

1.3.5 Definition 5

'Synthetic intermediate' means a chemical substance produced, stored, or used solely for chemical processing in order to be transformed into another chemical or other chemical substances.

Certain dangerous chemical agents are called carcinogenic, mutagenic, and toxic agents, classified in 'CMR' category, and consequently they are the subject of a particular treatment in terms of prevention. On the international level, labor codes include common provisions to DCA and their sub-category, CMR agents (Fig. 1.2).

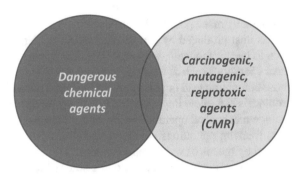

Fig. 1.2 Certain dangerous chemical agents are called carcinogenic, mutagenic, and toxic agents, classified in the 'CMR' category, and consequently they are the subject of a particular treatment in terms of prevention. This is a sub-category of the broad DCA group

1.3.6 Definition 6

A 'toxic substance' is a compound able to disrupt a living organism (Davis and Souza 2017). When an individual absorbs chemicals, a variety of biological effects may occur with beneficial (health improvement through a drug) or harmful (e.g., lung damage after inhalation of a corrosive gas) results. The concepts of toxic effects or intoxication involve harmful consequences for the body (United States Department of Labor 2013).

1.4   What Is a Harmful Effect on Health?

A general definition of an adverse health effect should be linked to visible bodily injury, diseases, modified growth or body development, effects on the developing fetus (teratogenic effects and fetotoxic effects), alterations in children, grandchildren, etc. (hereditary genetic effects), decrease in service life, changes in mental state associated with stress, trauma, exposure to solvents, etc., and finally alterations on the ability to cope with additional stress.

1.5   What Is a CMR Agent?

Chemicals or preparations with adverse health effects are rated 'CMR'. This ranking can come from European levels or other systems. These rankings are regularly updated according to the evolution of knowledge and products subjected to regulations in the field of work (use, protection, and supervision of the worker). Ideally, these products should be replaced by other 'less dangerous' ones: This is called 'substitution.'

CMR agents are defined as substances or preparations (chemicals) that are carcinogenic, mutagenic (genotoxic), and reproductive (toxic for reproduction), as displayed in Fig. 1.2, namely (European Chemical Agency 2012; Kasbi-Benassouli et al. 2005):

(1) Carcinogens (C). Dangerous chemical agents in its pure state (asbestos, wood dust, benzene, etc.) or in a mixture or process which, by inhalation, ingestion, or skin penetration, can cause cancer or increase its frequency. Wood dust can be an important agent in food production (smoking operations)
(2) Mutagenic (M) or genotoxic compounds. Substances or preparations which, by inhalation, ingestion, or skin penetration, induce alterations in the structure or number of chromosomes of the cells. Chromosomes are the elements of the nucleus of the cell that carry the genetic information. The mutagenic effect (or genotoxic damage) is an initial stage of cancer development. They can cause one or more hereditary genetic modifications or increase their frequency
(3) Toxic for reproduction or reprotoxic (R). Substances and preparations (e.g., lead) which, by inhalation, skin penetration, or other ways, may produce or increase the frequency of non-hereditary harmful effects in the offspring or impair reproductive function or ability. Specifically, these agents can alter human fertility and the development of the unborn child (spontaneous abortion, cause malformations in the fetus, etc.).

According to the European Regulation (EC) No 1272/2008, CMR is classified according to their dangerousness, the degree of knowledge, and certainty that one has on the substance or the preparation, in three categories:

(1) Category 1A: Those known to be carcinogenic, mutagenic, or toxic to reproduction, known as proven, that is to say that the effects in humans are recognized (e.g., benzene and lead compounds). These substances are known to be carcinogenic to humans. As an example related to canned foods, lead (a metal) is not allowed in metal cans for many years (contamination by packaging materials and objects)
(2) Category 1B: Those for which there is a strong presumption that human exposure to such substances and preparations may cause cancer and hereditary genetic defects, increase their frequency, produce or increase the frequency of non-hereditary harmful effects in the offspring, or impair reproductive functions. These substances to be assimilated to carcinogenic substances for humans (strong presumption that the substance may cause cancer)
(3) Category 2: Those which appear to be of concern to humans because of possible carcinogenic, mutagenic, or toxic effects for reproduction but for which the information available is insufficient to classify them in Category 1B, which are considered to be suspect. These substances are of concern for humans (possible carcinogens).

A preparation or mixture is classified as carcinogenic, mutagenic, or toxic for reproduction Category 1, 2, or 3 (in the repealed European Directives 67/548/EEC and 1999/45/EC) or 1A, 1B, or 2 in the new and in force CLP Regulation if it contains a component classified as a CMR agent at a concentration $\geq$ the concentration

Table 1.1 Classification of CMR substances in the European Union

Substance (classification)	According to repealed EU directives 67/548/EEC and 1999/45/EC		According to the EU CLP regulation	
	Category	Limit (in general) (%)	Category	Limit (%)
Cancerogen	1 and 2	≥0.1	1A and 1B	≥0.1
	3	≥1.0	2	≥1.0
Mutagen	1 and 2	≥0.1	1A and 1B	≥0.1
	3	≥1.0	2	≥1.0
Toxic for reproduction	1 and 2	≥0.5	1 and 2	≥0.3
	3	≥5.0	3	≥3.0
			Effect concerning or via breast-feeding	≥0.3

limit shown in Table 1.1 (except for substances whose classification threshold is specifically referenced).

Some notes should be considered. First of all, the substitution of the most dangerous chemical agents with less harmful substances or processes should be a priority in the prevention of chemical risks. According to the European Regulation (EC) No 1272/2008, substitution is an issue of the Occupational Health Plan published in 2005 and the new REACH Regulation (EC) No 1907/2006 (concerning the registration, evaluation, authorization, and restriction of chemicals), which covers in particular carcinogenic, mutagenic, and toxic for reproduction (CMR) substances. Thus, three categories of 'dangerous chemical agents' are not considered as CMR:

(1) Category 3 CMR (suspected effects)
(2) CMR classified as carcinogenic by the International Agency for Research on Cancer (IARC)
(3) CMR whose carcinogenic nature is recognized (e.g., crystalline silica dust).

Moreover, the occurrence of cancer depends on both the intensity of exposure to the product and the duration and frequency of contact with the carcinogen, and it also depends on individual factors. The use of chemicals—particularly CMR—requires prior consultation of complete product information sheets. Thus, the special rules of prevention to be taken against exposure risks to CMR agents can be summarized as follows:

(1) Substitution of CMR where technically possible. The results of the corresponding investigations are recorded in the single document
(2) Work in isolation (however, this practice in 'crowded' areas such as certain food industries may be difficult)
(3) Organizational and technical measures consisting of:

(3.1) Reduction of amounts
(3.2) Reduction of the number of employees exposed
(3.3) Capture at the source (extremely important in food environments where the compound is continually generated/produced)
(3.4) Implementation of hygiene measures
(3.5) Information and training for workers
(3.6) Area limitation of risk zones
(3.7) Setting up devices to be used in case of emergency
(3.8) Safe collection, storage, and disposal of waste
(3.9) Medical supervision.

Should CMR present other hazards, for example physicochemical dangers, the employer would put in place measures needed to suppress or reduce them (Brossat et al. 2012).

It has to be noted that some CMR agents have well-defined exposure limits: for example, benzene (3.25 mg/m^3); vinyl chloride monomer (2.59 mg/m^3); wood dust (1 mg/m^3); and metallic lead and its compounds (0.10 mg/m^3) (Cheng et al. 2000; European Commission 1991; Lewis 2008; Puntarić et al. 2005).

Another note concerns exposure to certain toxic chemicals (e.g., acrylic fibers and nylons) before the age of 36 and related possible risk of breast cancer after menopause, according to a Canadian study (Labrèche et al. 2010).

1.6 Elements of Occupational Toxicology

Occupational toxicology is an essential discipline concerning (i) primary prevention to avoid the occurrence of a chemical risk and (ii) secondary prevention in order to avoid further damages as soon as possible. Occupational toxicology must be developed within the framework of a multi-disciplinary chemical risk control: Harmful effects of chemicals, in terms of occupational health and also in terms of the environment, need to be pursued intensively. Epidemiological and toxicological studies concern approximately 100,000 chemicals, and this number is constantly growing while a very small number of them have been thoroughly tested for detecting their risks, especially carcinogens (Silbergeld 1998).

Occupational (or industrial) toxicology is used with the following aims:

(1) To describe biological effects of different industrial chemical agents used in the workplace
(2) To define permissible exposure levels and means for measuring the concentration of these substances in the air or on surfaces
(3) To define biological monitoring and screening for toxic effects in workers.

1.6.1 Toxicological Sheets

Toxicological records refer to pure substances. Their structure follows a seven-part standard plan (Ross and Godin 2016; Safe Work Australia 2012):

(1) Identification
(2) Features (uses, physical properties, chemical properties, storage containers)
(3) Fire and explosion
(4) Pathology and toxicology
(5) Regulations (occupational health and safety, environmental protection, protection of the population, transportation, etc.)
(6) Recommendations for storage, handling, and from a medical point of view
(7) Bibliography.

1.6.2 Occupational Exposure Limit Values

Occupational exposure limit (OEL) values are fixed concentration values of a hazardous chemical agent in the air that a person can tolerate for a specified time—during an eight-hour workday (Fig. 1.3) without health risks, even if physiological reversible changes may be sometimes tolerated (Japan Society for Occupational Health 2016). These limits aim to protect workers against adverse effects related to medium- and long-term regular exposure and for the duration of a working life to the chemical agent considered. OEL is usually expressed in volume (ppm or part per million) or in weight (mg/m^3).

OEL is a significant tool for risk assessment and management providing valuable information for occupational safety and health activities related to hazardous substances. Employers must ensure that the exposure of their workers does not exceed national limits.

Anyway, limit values are based on information on toxic properties of substances. Information is drawn from industrial experience (isolated observations, epidemiological surveys concerning immediate or long-term toxic effects) and studies on laboratory animals. Interestingly, the amount of studies concerning the use of selected chemicals (and related OEL values) in food and beverage industries has increased in the last year, demonstrating that a chemical risk exists in this ambit. However, workers are not immune to other effects, such as allergy phenomena. These values have three technical categories (AISS 2016; Picot and Grenouillet 1994; Ziem and Castleman 1989):

(1) The threshold limit value[1] (TLV®) or short-term exposure limit (STEL) which applies for periods ≤15 min. It corresponds to an exposure measured over a

---

[1]TLV® is a trademark of the American Conference of Governmental Industrial Hygienists (ACGIH®), https://www.acgih.org/.

Exposure (ppm or mg/m3)

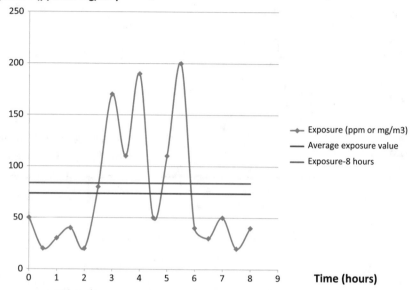

Fig. 1.3 OEL values—usually expressed in volume or in weight—are fixed concentration values of a hazardous chemical agent in the air that a person can tolerate for a specified time—during an eight-hour workday without health risks. OEL is a significant tool for risk assessment and management providing valuable information for occupational safety and health activities related to hazardous substances. The blue line means the real exposure, while purple and red lines concern the average of all recorded values and the recommended exposure in the same period, respectively

reference period of 15 min during a peak of exposure, whatever its duration. It aims to protect workers from immediate or short-term adverse effects due to exposure at concentrations above eight-hour OEL to occur during short periods of time during a working day (Fig. 1.4)

(2) The maximum or 'plafond' value or the concentration in the air of workplaces that must not be surpassed at any time during the day. This value is intended to protect against irritant effects that are strong and corrosive or that can cause a serious, potentially irreversible, effect in the very short term or immediately, as shown in Fig. 1.5

(3) The average exposure value (AEV) or also threshold limit value–time-weighed average (TLV-TWA), which is an eight-hour weighted average, and it can also be exceeded temporarily provided that it is ≤TLV (when available).

Exposure (ppm or mg/m3)

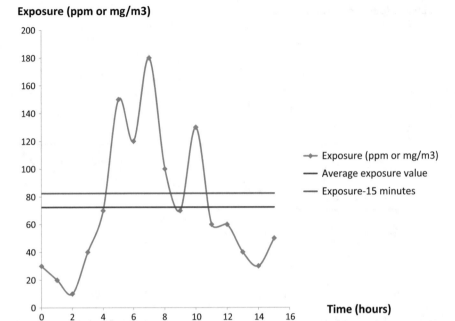

Fig. 1.4 The threshold limit value (TLV®) or short-term exposure limit (STEL) which applies for periods ≤15 min corresponds to an exposure measured over a reference period of 15 min during a peak of exposure, whatever its duration. It aims to protect workers from immediate or short-term adverse effects due to exposure at concentrations above eight-hour OEL to occur during short periods of time during a working day

1.6.3 Individual Factors

Men are not equal when facing chemical exposure. Individual factors should be considered. Many individuals may be affected differently by an identical toxic dose. The same person may also show a different reaction depending on the moment. Some factors explain the extent of toxic effects such as genetic differences or physiological factors:

(1) Age. Susceptibility to toxic effects is usually greater in children and the elderly than in other groups
(2) Sex. Absorption and metabolism are different for men and women
(3) Nutritional status. Toxicity can be influenced by adipose tissues, dehydration, alcohol, tobacco, etc.
(4) Pregnancy. Changes in the metabolic activity of the toxins occur during this period
(5) Health status. In general, healthy individuals are more resistant, than people suffering from liver or kidney diseases (two simple examples).

Exposure (ppm or mg/m3)

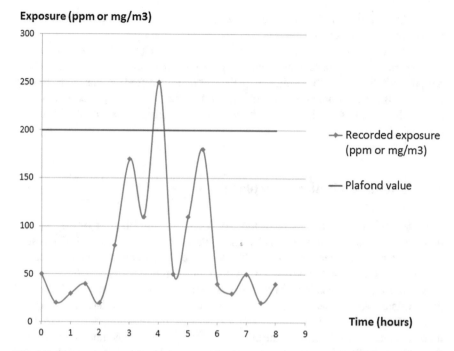

Fig. 1.5 The maximum or 'plafond' value is the concentration in the air of workplaces that must not be surpassed at any time during the day. This value is intended to protect against irritant effects that are strong and corrosive or that can cause a serious, potentially irreversible, effect in the very short term or immediately

Once the body barrier is crossed, the product diffuses into the body via the blood system. It can then reach all organs and cause various disorders in the body.

Target organs depend on the toxic substance (Hellman 2003; Lee et al. 2017). A product may preferentially affect heart, lungs, liver, kidney, gonads, eye, ear, etc. Neurotoxicants affect the nervous system. The term 'neurotoxic' does not mean that the product reaches only one organ, because other targets in the organism are possible.

In general, most solvents concentrate in fatty tissues such as the nervous system by disrupting its functions; all solvents are neurotoxic (encephalopathies). The inhalation of fine dust-crystallized silica or asbestos fibers has to be considered. An example useful in food environment is the possible use of low-molecular-weight silica powders for cheese productions (as anti-agglomeration agents) when speaking of sliced cheeses (Fox et al. 2000). These products enter the lungs and, by attaching themselves to the muscles, cause fibrosis (hardening of the lung muscles) resulting in a substantial decrease in respiratory capacity.

In addition, toxic products undergo complex chemical transformations called 'metabolism'; formed metabolites are often reactive, causing disturbances. These metabolites are at the origin of many occupational diseases. Toxic products, whether metabolized or not, either bind to organs or are eliminated in urine (kidneys), feces (intestines), or exhaled air (lungs).

It should also be noted that OEL differs for the same chemical depending on the country. In addition, there is one OEL only for one chemical in one country. These national differences can be explained by the fact that interested nations are following different OEL development processes. In addition, the OEL setting incorporates not only technical and scientific criteria but also social, political, and economic criteria (DGUV 2020).

1.7 Chemicals in the Workplace

Toxicity risks arise primarily from the physicochemical properties of the peculiar products (molecule and/or physical form). The chemical composition of the substance is often decisive, but harmful effects on one or more physiological functions do not depend only on the molecular structure of the product.

Silica, for example, is inert, but it becomes dangerous only when it is inhaled into fine particles. As an example, silica powder can be used as an antistacking agent in certain sliced cheeses under modified atmosphere. In this situation, particle sizes can mainly determine the toxicity—the entering is by ingestion or by inhalation (Merget et al. 2002). Similarly for fibers, for example asbestos, the risk is more related to the physical structure of the fiber than to its chemical structure.

Depending on the nature of the occupational activities and occupational hygiene behaviors, workers may be exposed to chemicals through several access routes:

1. Inhalation by inhalation to the pulmonary alveoli
2. Skin contact and penetration (such as certain alkaline detergents for sanitization purposes in food industries)
3. Oral ingestion and swallowing.

Chemicals can take different physical forms: solid (particles and dust), liquids (including fogs), gaseous (including vapors), and mixed (including fumes). The severity of exposure to toxic risks depends on (Hellman 2003; Stellman 1998):

(1) The intrinsic toxicity of the concerned chemical, tending to increase with hydrocarbon number, length and thickness of fibers, the reduced dimension of dust particles, etc.
(2) The chemical family (aromatic hydrocarbons and alcohols)
(3) The volatility. The lightest and the most volatile is the compound, the higher the supposed toxicity
(4) Concentration, frequency, localization (localized, systemic events), and duration (acute and chronic) of exposure
(5) The route of exposure (respiratory, cutaneous, ocular, or digestive options)
(6) The nature of toxic effects (irritant, sensitizing, asphyxiating, carcinogenic options)

(7) The presence in particular products (pesticides, soaps, food additives, solvents)
(8) Individual sensitivity (especially to allergens, and this aspect is particularly relevant to food and beverage-related industries
(9) Action mechanisms (stimulant and inhibitor compounds).

The classification of the toxic substance should be considered from several viewpoints. It often depends on the field of application, the objective pursued by an organization, and/or even the anthropic activity itself.

1.8 Toxicity of the Chemical

Toxicology is the study of adverse effects of chemical substances on living organisms (Hellman 2003; Stellman 1998). The toxicity of a foreign chemical (xenobiotic) to the organism is a biological characteristic that depends on the atomic or molecular structure of the compound and therefore its interaction with living matter. It is a measure of the poisoning capacity of a chemical.

This toxicity also depends on the dose of xenobiotic needed to produce a measurable effect. In industrial environments, the toxic effect depends on the composition of the air, the nature and physicochemical properties of the suspect substance, its conditions of use, and the individual himself. Industrial toxicology aims to determine a threshold of safety requiring detailed studies to identify changes and alterations of all kinds.

The main basis for the classification of chemicals is the assessment of exposure levels with the impact on the environment (water, air, and soil). About 50% of the international systems contain criteria relating to the volume of production of a chemical or the effects of pollutant releases. The most widely used criteria are mean 'lethal dose 50' (LD_{50}) and mean 'lethal concentration' (LC_{50}) values. These data are evaluated in laboratory animals according to three main routes of administration—oral, dermal, and inhalation options—for a single exposure. LD_{50} and LC_{50} are evaluated in the same animal species and using the same exposure routes (Hellman 2003; Stellman 1998).

Toxicologists often test animals to determine whether small or large doses of a specific chemical cause toxicity. One of these tests measures the chemical dose (LD_{50}) required to cause death in 50% of the animals. The greater the harmful effects of a xenobiotic, or the lower the threshold dose of this compound, the more toxic this compound is.

Depending on exposure conditions, most chemicals can cause both acute and chronic toxicities. Harmful effects of both kinds of toxicity are often very different. Toxicity is a measure of intoxication capacity and is an invariable characteristic of a chemical (Hellman 2003; Stellman 1998). Basically, the intoxication can be:

(1) Accidental, which most often consists of the absorption of a single massive dose, leading to immediate health effects such as irritation, burns, respiratory or ocular disorders

(2) Chronic, often due to the repeated administration of small doses, resulting in the possible appearance over time of an occupational disease (e.g., occupational cancer).

In addition, the risk of fire explosion has to be considered. Dangerous chemical reactions likely to produce dangerous, toxic, and/or flammable substances, thermal burns, and asphyxiation are all possible consequences. A highly toxic chemical can be a low health hazard if used with proper precautions and measures. On the other hand, a chemical with low toxicity presents a serious health hazard if used with no care.

All chemicals can cause poisoning if the dose is sufficient. In other words, all chemicals can be toxic. Meanwhile, chemicals that are only slightly toxic should be absorbed in large doses to cause intoxication. Small doses of highly toxic chemicals are enough to cause poisoning. Anyway, the quantity or the assimilated dose by the organism determines whether toxic effects will be observed. So, poisoning is caused not only by exposure to a specific chemical, but rather by exposure to extremely abundant doses of product. It should also be noted that the toxicity of a chemical cannot be modified, while hazards can be managed and adequately minimized. Two types of toxicity are to be distinguished: acute and chronic toxicities (Hellman 2003; Stellman 1998).

1.8.1 The Toxic Effect

The toxic effect is the consequence of the accidental release of polluting product(s) in the form of gas clouds, following, e.g., a rupture of piping, the destruction of storage tanks or a fire, or the opening of a melting reactor containing an intermediate food mass with dispersed citric acid (with consequent aerosolized compounds). Should an individual absorb chemicals, various biological effects would occur and appear:

(1) Beneficial, e.g., improvement of health after administration of a medication, or
(2) Harmful, e.g., lung damage following the inhalation of a corrosive gas.

The notion of toxic effect implies harmful consequences for the organism (Hellman 2003; Stellman 1998). Inhaling, touching, and even ingesting chemicals do not necessarily result in a toxic effect. For example, carbon dioxide (CO_2) is a metabolite in the human body exhaled by the lungs that is also found in the environment. It causes suffocation if present in sufficient quantity in an enclosed or poorly ventilated space. Paradoxically, the absorption of a substance in small quantity can prove to be very toxic and cause serious lesions, while the large absorption of another substance with lower toxicity can produce a favorable effect. Therefore, the harmful effect is related to the dose, the absorption route, the type, and severity of the lesions as well as the time necessary for the appearance of a lesion. The toxic effect is thus linked to the notion of toxicity.

For each product, three thresholds of effects are identified with regard to its harmfulness (Hellman 2003; Stellman 1998):

Table 1.2 Examples of toxic effects on certain biological tissues and systems

Targeted system and related organs	Clinical effect or sign
Eye	Irritation and corrosion
Skin	Irritation, corrosion, dermatosis
Digestive system	Irritation and corrosion
Cardiovascular system	Abnormal heart rate
Central nervous system	Depression (nausea, vomiting, etc.)
Peripheral nervous system	Neuropathy
Respiratory system	Irritation, corrosion, dyspnea (shortness of breath)
Blood system	Carboxyhemoglobinemia
Urinary system	Very dark urine and blood in the urines

(1) The threshold of irreversible effects (concentration beyond which permanent sequelae could appear on people)
(2) The threshold of the first lethal effects (concentration beyond which deaths could be observed in 1% of the population)
(3) The threshold of significant lethal effects (concentration threshold above which deaths could be observed in 5% of the population).

Toxicity encompasses all of the harmful effects of a toxin on a specified life form. In other words, it is the inherent capacity of a chemical to produce harmful effects in a living organism (Table 1.2).

A toxic cloud released into the atmosphere can spread, move under the effects of weather conditions (wind and stability of the atmosphere in particular), and ultimately dissipate as soon as the leak is controlled. The toxic cloud will have an effect on the human person if it reaches the environment where he/she is located. The toxic effect on humans depends on the product, its concentration in the air, exposure times, and wind flows in closed or open environments, such as food and beverage industries (the possible presence of openings and leaks in constructions into buildings has to be considered).

The toxic effect is the result of often complex process(es), and it can lead to a series of physiological and metabolic reactions (e.g., irritant gas on the respiratory system). The unfortunate thing is that some toxic gases do not smell, and most are not seen.

1.8.1.1 Acute Toxicity

In relation to acute toxicity (caused by massive assumptions of the xenobiotic agent, in other terms: poisoning), related effects are immediate. In most cases, much more is known about the acute toxicity of a chemical than about its chronic toxicity. Our understanding of acute toxicity usually comes from studies with animals that have been exposed to relatively high doses. Accidental overexposure, spills, and emergencies have allowed us to learn more about acute toxicity in humans. Health effects can be temporary, such as skin irritation, illness, or nausea, or permanent: blindness, scarring from acid burns, reduced intellectual capacity, etc. (Hellman 2003; Silbergeld 1998; Stellman 1998).

Acute toxicity often is observed within minutes or hours after sudden and intense exposure to a chemical. However, effects of exposure to high concentrations may appear later. For example, symptoms associated with intense exposure to certain pesticides may not appear for several days.

1.8.1.2   Chronic Toxicity

Chronic or long-term toxicity has been studied so far in animals. In addition, we have also learned a lot from studies of groups of people exposed in the workplace to chemicals for years (Hellman 2003; Silbergeld 1998; Stellman 1998). Several studies also concern food operators (Damalas and Koutroubas 2016).

As a general rule, chronic toxicity appears many years after the initial exposure. The disease develops only because there has been repeated exposure over many years. Diseases of chronic toxicity do not appear to be caused by single, sudden exposure. Chronic toxicity is believed to involve either of the following two mechanisms, which can be illustrated using sodium fluoride and n-hexane as examples:

(1) Sodium fluoride, in very low concentrations (as in toothpaste or drinking water), does not cause visible harmful effects, even after years of exposure. Even at these low concentrations, effects are positive for teeth. However, when the body is repeatedly exposed to much higher concentrations, sodium fluoride is deposited and accumulated in the bones. At first, the amount of fluoride is not a problem, but after years of repeated exposure to high doses, there may be symptoms of bone disease

(2) n-hexane does not deposit and does not accumulate in the body. It degrades in the liver. A breakdown product can 'attack' nerve cells in the fingers and toes. These cells do not regenerate easily. Over the years, continuous exposure increases cell damage to such an extent that symptoms occur in the nerves of the fingers and toes.

A special case of chronic toxicity is cancer. Repeated exposure to certain chemicals for long time periods can cause cancer. Although there is no compelling evidence that cancer can develop after a single exposure, most cases suggest that repeated exposure over a long period is necessary (Kasbi-Benassouli et al. 2005).

Moreover, there is a tendency to believe that if small amounts of a chemical are enough to cause intoxication, the chemical is in principle very dangerous. This is not necessarily the case.

1.9 Why Is a Chemical Toxic?

Dangerous chemicals for human health must first come into contact with, or enter that person's body, and must have a biological effect on the body itself. As a result, it is necessary that (i) the organism is exposed to a toxic compound, that (ii) this toxic compound enters it, and that (iii) the organism absorbs a sufficient quantity able to disturb its functioning. It should be noted that the majority of toxins must generally enter the organism to produce harmful effects, except those causing local effects.

The main four routes of entry are (Hellman 2003; Silbergeld 1998; Stellman 1998):

- Inhalation (breathing)
- Skin contact
- The digestive system (ingestion or diet)
- Injections.

With reference to food and beverage environments, the most probable routes should exclude the ingestion and injections.

It has to be remembered that organisms operate under relatively constant conditions (pH, oxygen, etc.). This situation is called 'homeostasis,' and living organisms seek to maintain this balance. The human body can be adaptable to many aggressions, whether biological, physical, or chemical. The body's adaptation processes work continuously to ensure that this balance is maintained (Fig. 1.6). When this

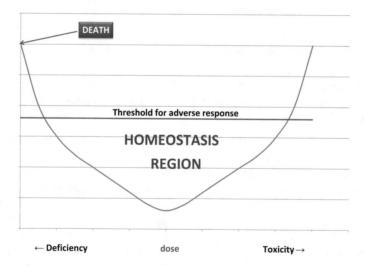

Fig. 1.6 Organisms operate under relatively constant conditions (pH, oxygen, etc.). This situation is called 'homeostasis,' and living organisms seek to maintain this balance. The human body can be adaptable to many aggressions. The adaptation processes work continuously to ensure that this balance is maintained. When this equilibrium is modified, it causes a dysfunction (toxic effect, right)

equilibrium is modified, it causes a dysfunction (toxic effect). There is mobilization of a part of the organism and sometimes of the entire organism; various reactions are triggered with the aim of counterbalancing the attack and restoring the broken balance. From now on, the organism can resist a toxic aggression as long as it is carried out inside the limits of its mechanisms of detoxication, homeostasis, and repair. Beyond this, compensation mechanisms can be insufficient (Davies 2016; Kotas and Medzhitov 2015). The defense system cannot then counteract toxic effects.

It should be noted that the toxic substances do not produce effects of the same intensity on all the organs (e.g., the kidney) or the tissues (e.g., the blood). They attack organs, in particular selected 'target' organs, for reasons that are not always understood. There can be several reasons, including a greater sensitivity of these organs, a higher concentration of the toxic compound and/or its metabolites, etc. For example, the liver is a target organ for vinyl chloride. As a result, for prevention, the following factors should be studied (Hellman 2003; Silbergeld 1998; Stellman 1998):

(1) The nature and composition of used products
(2) The toxicity of these products and the thresholds for the appearance of this toxicity
(3) Detection methods in the atmosphere
(4) Screening methods for exposure, both clinically and biologically
(5) Collective and individual means of prevention and their degree of effectiveness and acceptability.

1.10 Prevention Approaches

Two approaches are possible to prevent health risks from the presence of chemicals in the workplace, namely (Hellman 2003; Ross and Godin 2016; Silbergeld 1998; Stellman 1998):

(1) The establishment of limit values for the concentration of substances in the air of industrial atmospheres (including food and beverage industries), that is to say a concentration which does not cause any toxic effect in the exposed subjects, in the case of substances which enter the body by inhalation. Environmental air analysis will therefore make it possible to control whether working conditions are permissible, that is to say in fact if the prevention methods are respected. Naturally, the removal of 'exhausted' atmosphere from production/storage areas in food industries and the replacement with 'fresh' air are good systems for limiting exposure, but related effects have to be evaluated after initial measurements
(2) Analysis of one or more biological media (blood, urine, expired air, etc.) of the exposed subjects, by which we try to estimate either the total exposure of the individual to a chemical compound (by measurement of concentrations

of the compound itself or one of its metabolites) and the intensity of a biological response that has not yet produced a pathological manifestation. This
second strategy, related to the quantity of absorbed product, will be developed
elsewhere.

1.11 Factors Influencing Toxicity Degrees

Several factors can influence the degree of toxicity of a chemical, namely (Hellman
2003; Silbergeld 1998; Stellman 1998):

1. Pathway into the body
2. Quantity (dose) entering the body
3. Toxicity of the chemical
4. Elimination of the organism
5. Biological variations.

Some factors are or appear extremely important. Two examples can be made here.

Does the amount or dose entering the body matter? The dose of a chemical entering
the body is probably the single most important factor to determine if a chemical
product will cause poisoning. The amount that can cause intoxication depends on
the chemical.

In addition, the elimination rate of the body may be important. Many chemicals
present in the workplace enter the body and are unchanged when they are eliminated,
while others are degraded. The degradation products may be more toxic or less toxic
than the original chemical. Other chemicals are stored temporarily in the organs and
are eliminated soon after. Ultimately, the most part of chemicals and their breakdown
products are eliminated in the stool, urine, sweat, and expired air. Some chemicals,
such as silica or graphite dust, can be inhaled, lodged in the lungs, and left there for
many years. They may never be completely eliminated.

Consequently, the risk of illness caused by a chemical is lower if the body is able
to do one or both of the following (excretion) actions:

1. To degrade the chemical into less toxic products
2. To quickly remove the chemical from the body.

1.12 Main Dangers in General

The real chemical risk is invisible and insidious: Related effects can be delayed in time
(cancers) and sometimes without links with those who use them (bioaccumulable
products). The clear and precise regulations set anchor points for a good control of
the chemical risk. Far from being expensive and insurmountable, it only takes a little

rigor and complies with a few essential rules (Dikshith 2013; Gorguner and Akgun 2010):

(1) Contact with concentrated acids and bases may cause severe burns of skin, mucous membranes, and eyes
(2) The extent of the damage depends on four factors: the corrosion power, the concentration, the temperature of the solution, and contact duration
(3) Inhalation of acid vapors may cause irritation and burns of the respiratory tract
(4) Chronic exposure to weak or diluted acids and bases can cause tissue damage (dermatitis). Salicylic acid or oxalic acid may, for example, reduce the keratin layer protecting the skin with consequent tissue irritation
(5) In addition to direct tissue damage, some acids, such as chromic acid or picric acid, even if diluted, may cause allergic reactions.
(6) The dilution of acids/bases concentrated in water is strongly exothermic and may lead to splashing.

1.13 Endocrine Disruptors

An endocrine disruptor is, according to the WHO, a chemical substance of natural or synthetic origin, able to interfere with the endocrine system, that is to say cells and organs involved in the production of hormones and their action on so-called target cells. These substances are suspected of causing malformations, male infertility problems, or even increasing the risks of developing certain cancers with additional allergic reactions (Bergman et al. 2013; WHO 2020).

In general, endocrine disruptors are capable of acting on the synthesis, secretion, transport, binding, or elimination of natural hormones. However, these hormones are involved in reproduction, growth and development, maintenance of homeostasis (maintenance of the internal environment), and energy availability. As a result, an endocrine disruptor is responsible for possible disturbances in all of these functions, causing harmful effects on the health of an organism or its descendants, by acting on the endocrine system. As a reminder, the role of the endocrine system is to regulate many of the complex activities of the human body.

The most important and well-documented health concerns related to endocrine disruptors are effects on development and reproductive effects (Bergman et al. 2013; WHO 2020). Some of the disorders that have been observed in animal studies include the following: oligospermia (low sperm count), testicular cancer, and hyperplasia of the prostate in adult males and adenocarcinoma of the vagina, ovulation disorders, breast cancer, and uterine fibroids in adult females. Disruption of thyroid function, obesity, bone metabolism, and diabetes are also linked to exposure to endocrine disruptors.

One of the characteristics of endocrine disruptors is the absence of a dose–response relationship. Indeed, for a 'conventional' pollutant, the more the pollutant is present in large quantities in an organism, the more toxic it is. On the contrary, this relationship is not valid for endocrine disruptor because the compound acts like a hormone. Thus, effects can be highlighted at minute concentrations than a reduction of this effect at a higher concentration. This property makes the effect of endocrine disruptors difficult to predict (Bergman et al. 2013; Hellman 2003; WHO 2020).

Endocrine disruptors[2] are extremely varied in nature:

(1) Industrial chemical substances: plasticizers such as phthalates, bisphenol A or alkylphenols, perfluorinated compounds, etc. As a simple example of interest, bisphenol A is related to metal can coatings and other plastic-made objects for food purposes (packaging) (Parisi 2012, 2013)
(2) Pesticides: organochlorines, vinclozine, etc. Clearly, their presence in harvested products is a high concern
(3) Aromatic hydrocarbons: polychlorinated biphenyls (PCBs), etc. Their presence in certain materials such as fuel for thermal reactors and burned wood (for smoking operations) is a food-related concern
(4) Medicines: anticancer drugs such as tamoxiphene, natural and synthetic estrogens (contraceptive pill), etc. The possible presence of unallowed medicines in animal tissues has to be considered
(5) Natural substances: phytoestrogens, isoflavonoids, etc.

How does an endocrine disruptor work? It acts according to three modes of action:

(1) Mimetic effect: It can imitate the action of natural hormones such as estrogen or testosterone, like a false key in the 'biological locks' that exist in organs and cells
(2) 'Locking' effect: By blocking the receptors of cells receiving hormones (hormone receptors), thus preventing the action of hormones (by saturating the receptors, for example)
(3) 'Disturbing' effect: By acting on the synthesis, transport, metabolism, and excretion of hormones, thus modifying the concentrations of natural hormones. These disturbances are all the more serious as they occur early (fetus, embryo, young) because irreversible effects can be induced.

1.14 Chemicals and Dangers: A Deep Analysis

Chemical hazards are ubiquitous in human activities and are not specific to chemical or para-chemical industries (such as food and beverage companies): construction, cleaning, health, personal services, metal exploitation, textile products, furniture, food industry, carpentry, transport, and waste treatment (Hellman 2003; Silbergeld 1998; Stellman 1998). Chemicals present dangers to people, installations, and/or the environment: acute intoxications, asphyxia, fire, explosion, pollution, ototoxic, etc.

---

[2]An interested and constantly updated list can be found at the following Web address (by the European Chemicals Agency): https://echa.europa.eu/it/ed-assessment.

They can also cause more insidious effects, after years of worker exposure at low doses or even years after the end of exposure. These immediate and delayed hazards must be taken into account as part of the same approach to chemical risk prevention.

Any chemical that comes into contact with the human body can lead to more or less serious health effects. Whether solid, liquid, or gaseous, chemicals use three main routes to enter the body: by inhalation, in contact with the skin, or by ingestion. Two types of exposure are distinguished in a workplace, namely (Hellman 2003; Silbergeld 1998; Stellman 1998):

(1)   The products are used deliberately, in their liquid, solid, or gaseous state, for their properties (diluent, degreaser, etc.), or as intermediates with other products to manufacture a material or another substance, and this under specific conditions of implementation (application with a cloth or roller, by soaking, by spraying, at high temperatures, under pressure, etc.). This use may give rise to exposures

(2)   A process or activity gives rise to emissions of chemicals (dust, vapors, gases, fumes, mists, etc.): There is pollution of the workstation or its environment, hence possible exposure of the operator or company employees. All sectors of activity are affected by this type of exposure.

Occupational exposure may be related to a normal situation or function if preventive measures are inadequate, inappropriate, or not applied. Exposure can also be accidental: rupture, leak, poorly controlled process, accidental spillage, etc. (Norskolje&gass 2018).

Note that inhalation is the most common occupational mode of exposure, above all in food and beverage environments where notable amounts of powders for food purposes are used (as food additives). Next comes the percutaneous route, by contact and/or by diffusion through the skin: The effects are then either local (irritation, burning, necrosis, etc.) or general ones. The mode of exposure by ingestion is not negligible in the professional environment either because of ingestion by swallowing of previously inhaled substances (by swallowing saliva) or because of hygiene problems (dirty hands). The use of certain detergents and disinfection agents is a good example for this exposure (Gurnari 2015; Hellman 2003; Silbergeld 1998; Stellman 1998). Ingestion may also result from a voluntary act (suicide) or be accidental.

1.14.1   Danger Categories

Substances and preparations corresponding to the following categories may be classified as dangerous substances, according to the pre-CLP Regulation in Europe, as shown in Fig. 3.1 (Hellman 2003; Silbergeld 1998; Stellman 1998):

(1)   Explosives: solid, liquid, pasty, or gelatinous substances and preparations which, even without the intervention of atmospheric oxygen, may exhibit an exothermic reaction with rapid gas development and which, under certain conditions, detonate, deflagrate rapidly, or, under the effect of heat, explode in case

of partial confinement. Clearly, such a danger cannot be excluded in certain food areas

(2) Oxidizing agents: substances and preparations which, in contact with other substances, in particular flammable compounds, exhibit a strongly exothermic reaction

(3) Extremely flammable agents: liquid substances and preparations with extremely low flash point and low boiling point, as well as gaseous substances and preparations which, at ambient temperature and pressure, are flammable in air

(4) Highly flammable substances and preparations which:

(4.1) May be heated to the point of igniting in air at room temperature without the addition of energy

(4.2) May be easily ignited in the solid state by a brief action of an ignition source and continue to burn or burn after the source has been removed

(4.3) Show very low flash point in the liquid state

(4.4) Or, in contact with water or humid air, produce extremely flammable gases in dangerous quantities

(5) Flammable compounds: liquid substances and preparations with a low flash point

(6) Very toxic agents: substances and preparations which, by inhalation, ingestion, or skin penetration in very small quantities, cause death or cause acute or chronic health damage

(7) Toxic substances and substances which, by inhalation, ingestion, or skin penetration in small quantities, cause death or harm to health acutely or chronically

(8) Harmful agents: substances and preparations which, by inhalation, ingestion, or cutaneous penetration, can cause death or harm health acutely or chronically

(9) Corrosive substances and preparations which, in contact with living tissue, can exert a destructive action on the latter

(10) Irritant agents: non-corrosive substances and preparations which, through immediate, prolonged, or repeated contact with the skin or mucous membranes, can cause an inflammatory reaction

(11) Sensitizers: substances and preparations which, by inhalation or skin penetration, may give rise to a hyper-sensitization reaction such that subsequent exposure to the substance or preparation produces characteristic harmful effects

(12) Carcinogens: substances and preparations which, by inhalation, ingestion, or skin penetration, can cause cancer or increase its frequency:

(12.1) Category 1 carcinogens: substances and preparations known to be carcinogenic to humans

(12.2) Category 2 carcinogens: substances and preparations for which there is a strong presumption that human exposure to such substances and preparations may cause cancer or increase its frequency

(12.3) Category 3 carcinogens: substances and preparations which are of concern to humans because of possible carcinogenic effects, but for which insufficient information is available to classify these substances and preparations in category 2

(13) Mutagens: substances and preparations which, by inhalation or skin penetration, can produce hereditary genetic defects or increase their frequency:

(13.1) Category 1 mutagens: substances and preparations known to be mutagenic to humans

(13.2) Category 2 mutagens: substances and preparations for which there is a strong presumption that exposure of humans to such substances and preparations may produce heritable genetic defects or increase their frequency

(13.3) Category 3 mutagens: substances and preparations of concern to humans because of possible mutagenic effects, but for which insufficient information is available to classify these substances and preparations in Category 2

(14) Toxic compounds for reproduction: substances and preparations which, by inhalation, ingestion, or skin penetration, can produce or increase the frequency of non-hereditary harmful effects in the offspring or impair their productive functions or capacities:

(14.1) Category 1 reproductive toxicants: substances and preparations known to be toxic to human reproduction

(14.2) Category 2 reproductive toxicants: substances and preparations for which there is a strong presumption that human exposure to such substances and preparations can produce or increase the frequency of non-hereditary harmful effects in the offspring or undermine reproductive functions or capabilities

(14.3) Category 3 reproductive toxicants: substances and preparations of concern due to possible reproductive toxicity, but for which insufficient information is available to classify these substances and preparations as Category 2

(15) Dangerous compounds for the environment: substances and preparations which, if they entered the environment, would present or could present an immediate or deferred risk for one or more of its components.

1.14.2   A Special Category. Fire and Explosion Hazards

Chemicals can play a role in starting a fire through their presence in air or when mixed with other products. They can also aggravate the extent of a fire event. Many substances can also cause explosions of varying size and severity if they are flammable, combustible, or unstable products, under certain conditions. They are mostly gases and vapors, but also flammable dust and particularly unstable compounds. In relation to 'crowded' areas such as certain food production areas (without a definite route for entering raw materials and packaging materials, and exiting final products, intermediates, and waste), similar events can occur (Davletshina and Cheremisinoff 1998).

1.14.3   Dangerous Chemical Reactions

Finally, the mixture of incompatible chemical agents, the heating of products, thermal degradation, friction, or even shocks can cause massive emissions of toxic vapors, exothermic phenomena resulting in a deflagration, a detonation, splashes of material or inflammation, etc.

1.15   Other Hazards: A Short Selection

1.15.1   Hydrofluoric Acid

Hydrofluoric acid (HF) is a deadly risk: It is not used in the food and beverage industry, but it is also discussed here as a simple example in relation to remarkable risks. It can indeed cross the dermis and attack the organism in depth (tissue necrosis). Its handling requires special precautionary measures for the skin. A burn with concentrated HF (50–70%) is immediately felt, whereas it can take several hours to feel a burn with diluted HF (1–25%). In addition, a HF burn on only 2% of the body can already prove to be lethal.

1.15.2   Acids and Bases in General

In addition to their corrosive nature, some acids are toxic for the body and therefore harmful by inhalation. Bases generally penetrate tissue deeper than acids. The damage linked to burns from basic solutions is therefore potentially greater (in particular for eyes). In the event of leakage, spillage, or handling errors, acids and bases may come into contact with incompatible products (e.g., organic solvent + nitric acid) and cause hazardous reactions.

Omnipresent in the workplace, chemicals sometimes go unnoticed. However, many chemicals can have effects on humans and their environment. Identifying dangerous chemical products, mixtures, or processes and their effects is a first step before implementing appropriate prevention measures.

From now on, poorly identified and, above all, poorly managed chemical risks can prove to be much more important than any other industrial risk both at a scale level and at a gravity level. Indeed, it is often the exposure to small doses repeated over time (chronic exposure) which proves to be the most disastrous for human health. Not to mention the so-called persistent or bioaccumulable products that can end up miles away from the source of contamination and can affect anyone at any time, the risk becomes difficult to identify.

The main driver for supporting people to work longer is now to improve the quality of work. Improving employment practices for older workers will improve the workplace for all.

One specific acid of normal use in the food industry is briefly discussed in Chap. 2: citric acid, as food additive, as a simple example of difficulties shown by a single chemical with more than one declared application and feature on the one side and more than one specific danger for food operators. It has to be considered that many acids such as lactic acid (Ameen and Caruso 2017) may be used with citric acid or replace it when speaking of similar functions, but with very different dangers.

References

AISS (2016) Valeurs limites d'exposition pour la prévention des agents chimiques. Association internationale de la sécurité sociale (AISS), Institut National de Recherche et de Securitè (INRS), Paris. Available http://www.inrs.fr/dms/inrs/CataloguePapier/ED/TI-ED-6254/ed6254.pdf. Accessed 10 Jan 2020

Ameen SM, Caruso G (2017) Lactic acid in the food industry. Springer International Publishing, Cham. https://doi.org/10.1007/978-3-319-58146-0

ATA (2016) Understanding the facts. Causes. American Tinnitus Association (ATA), Washington, DC. Available https://www.ata.org/understanding-facts/causes. Accessed 08 Jan 2020

Bédard S, Berger M, Gobeil M, Goulet S, Paillé M, Sicotte D, Thibeault T (2009) Techniques et équipements de travail en hygiene et salubrité. La Direction des communications du ministère de la Santé et des Services sociaux du Québec, Montréal. Available https://publications.msss.gouv.qc.ca/msss/fichiers/2009/09-209-01.pdf. Accessed 08 Jan 2020

Bergman Å, Heindel JJ, Jobling S, Kidd K, Zoeller RT (eds) (2013) State of the science of endocrine disrupting chemicals—2012. World Health Organization (WHO), Genéve, and the United Nations Environment Programme (UNEP), Nairobi. Available https://www.who.int/ceh/risks/cehemerging2/en/. Accessed 09 Jan 2020

Binetti R, Costamagna FM, Marcello I (2008) Exponential growth of new chemicals and evolution of information relevant to risk control. Annali Istituto Superiore di Sanità 44, 1:13–15. Available http://old.iss.it/binary/publ/cont/ANN_08_04%20Binetti.1209032191.pdf. Accessed 08 Jan 2020

Brossat C, Thomasset L, Cote D, Certain E (2012) 10 bonnes Pratiques favorisant la santé au travail en contribuant à la performance globale des PME; Expériences conduites en Rhône-Alpes. SP1177- May 2012. Carsat Rhône-Alpes, Direction des Risques Professionnels et de la Santé au Travail, Lyon cedex. Available https://www.carsat-ra.fr/images/pdf/entreprises/sp1177.pdf. Accessed 10 Jan 2020

CAS (2020) CAS Content—the world's largest collection of chemistry insights. chemical abstracts service (CAS), columbus. Available https://www.cas.org/about/cas-content. Accessed 08 Jan 2020

Cheng TJ, Chou PY, Huang ML, Du CL, Wong RH, Chen PC (2000) Increased lymphocyte sister chromatid exchange frequency in workers with exposure to low level of ethylene dichloride. Mutat Res 470:109–114. https://doi.org/10.1016/S1383-5742(00)00045-4

CNRS (2020) CMR—le Cancérogène-Mutagène- toxique pour la Reproduction. Chemical Risk Prevention Unit, Centre National de la Recherche Scientifique, Gif-sur-Yvette Cedex. Available http://www.prc.cnrs.fr/IMG/pdf/memo-A4-cmr-01-12.pdf. Accessed 08 Jan 2020

Codex Alimentarius (1995) General standard for food additives, CODEX STAN 192–1995. Codex alimentarius commission, last revision 2018. The Food and Agriculture Organizaion of the United Nations (FAO), Rome, and the World Health Organization (WHO), Geneva. Available http://www.fao.org/gsfaonline/docs/CXS_192e.pdf. Accessed 08 Jan 2020

Csuros M (1997) Environmental sampling and analysis: lab manual. CRC Press, Boca Raton

Damalas C, Koutroubas S (2016) Farmers' exposure to pesticides: toxicity types and ways of prevention. Toxics 4(1):1. https://doi.org/10.3390/toxics4010001

Davies KJ (2016) Adaptive homeostasis. Mol Aspects Med 49:1–7. https://doi.org/10.1016/j.mam.2016.04.007

Davis L, Souza K (2017) Occupational and environmental health surveillance. In: Levy BS, Wegman DH, Baron SL, Sokas RK (eds) Occupational and environmental health. Oxford University Press, Oxford, pp 101–114

Davletshina TA, Cheremisinoff NP (1998) Fire and explosion hazards handbook of industrial chemicals. Noyes Publications, Westwood

DGUV (2020) GESTIS—international limit values for chemical agents (Occupational exposure limits, OELs). German social accident insurance (DGUV), Berlin. Available https://www.dguv.de/ifa/gestis/gestis-internationale-grenzwerte-fuer-chemische-substanzen-limit-values-for-chemical-agents/index-2.jsp. Accessed 09 Jan 2020

Dikshith TSS (2013) Hazardous chemicals: safety management and global regulations. CRC Press, Boca Raton

European Chemical Agency (2012) CMR substances from Annex VI of the CLP Regulation registered under REACH and notified under CLP. A first screening—Report 2012. ECHA-12-R-01-EN. European Chemical Agency, Helsinki. Available https://echa.europa.eu/documents/10162/13562/cmr_report_en.pdf. Accessed 08 Jan 2020

European Commission (1991) Recommendation from the scientific committee on occupational exposure limits for benzene, SCOEL/SUM/140, December 1991. European Commission, employment, social affairs and inclusion, brussels. Available https://ec.europa.eu/social/BlobServlet?docId=7423&langId=en. Accessed 08 Jan 2020

European Commission (2013) Chemicals at work—a new labelling system. Guidance to help employers and workers to manage the transition to the new classification, labelling and packaging system. European Commission, Directorate-General for Employment, Social Affairs and Inclusion, Unit B3, Brussels. Available https://osha.europa.eu/it/file/49187/. Accessed 08 Jan 2020

Fox PF, Guinee TP, Cogan TM, McSweeney PLH (2000) Fundamentals of cheese science. Aspen Publishers, Gaithersburg

Gorguner M, Akgun M (2010) Acute Inhalation Injury. Eurasian J Med 42(1):28–35. https://doi.org/10.5152/eajm.2010.09

Gurnari G (2015) Safety protocols in the food industry and emerging concerns. Springer International Publishing, Cham. https://doi.org/10.1007/978-3-319-16492

Havet N, Penot A, Morelle M, Perrier L, Charbotel B, Fervers B (2017) Trends in occupational
    disparities for exposure to carcinogenic, mutagenic and reprotoxic chemicals in France 2003–10.
    Eur J Pub Health 27(3):425–432. https://doi.org/10.1093/eurpub/ckx036
Health and Safety Authority (2020) Dangerous chemicals. Health and safety authority, Dublin.
    Available   https://www.hsa.ie/eng/Your_Industry/Fishing/Hazards/Dangerous_Chemicals/.
    Accessed 08 Jan 2020
Hellman B (2003) Basic toxicology. In: Waldron HA, Edling C (eds) Occupational health practice.
    Butterworth-Heimemann, Oxford, pp 18–34
Japan Society for Occupational Health (2016) Recommendation of occupational exposure limits. J
    Occup Health 58(5):489–518
Kasbi-Benassouli V, Imbernon E, Iwatsubo Y, Buisson C, Goldberg M (2005) Confrontation des
    cancérogènes avérés en milieu de travail et des tableaux de maladies professionnelles. Institut de
    Veille Sanitaire, Saint-Maurice cedex. Available https://www.vie-publique.fr/sites/default/files/
    rapport/pdf/054000575.pdf. Accessed 10 Jan 2020
Kotas ME, Medzhitov R (2015) Homeostasis, inflammation, and disease susceptibility. Cell
    160(5):816–827. https://doi.org/10.1016/j.cell.2015.02.010
Labrèche F, Goldberg MS, Valois MF, Nadon L (2010) Postmenopausal breast cancer and
    occupational exposures. Occup Environ Med 67(4):263–269. https://doi.org/10.1136/oem.2009.
    049817
Laganà P, Campanella C, Patanè P, Cava MA, Parisi S, Gambuzza ME, Delia S, Coniglio MA
    (2019a) Food gases: classification and allowed uses. In: Chemistry and hygiene of food gases.
    Springer International Publishing, Cham
Laganà P, Campanella C, Patanè P, Cava MA, Parisi S, Gambuzza ME, Delia S, Coniglio MA
    (2019b) Food gases in the European Union: the legislation. In: Chemistry and hygiene of food
    gases. Springer International Publishing, Cham
Laganà P, Campanella C, Patanè P, Cava MA, Parisi S, Gambuzza ME, Delia S, Coniglio MA
    (2019c) Food gases in the industry: chemical and physical features. In: Chemistry and hygiene
    of food gases. Springer International Publishing, Cham
Laganà P, Campanella C, Patanè P, Cava MA, Parisi S, Gambuzza ME, Delia S, Coniglio MA
    (2019d) Safety evaluation and assessment of gases for food applications. In: Chemistry and
    hygiene of food gases. Springer International Publishing, Cham
Lee BM, Kacew S, Kim HS (2017) Lu's basic toxicology: fundamentals, target organs, and risk
    assessment. CRC Press, Boca Raton
Lewis RJ (2008) Hazardous chemicals desk reference. Wiley, Hoboken
Merget R, Bauer T, Küpper H, Philippou S, Bauer H, Breitstadt R, Bruening T (2002) Health hazards
    due to the inhalation of amorphous silica. Arch Toxicol 75(11–12):625–634. https://doi.org/10.
    1007/s002040100266
Norskolje&gass (2018) Chemical exposure in connection with accidents and spills. Norskolje&gass,
    Stavanger. Available https://www.norskoljeoggass.no/en/operations/storulykkerisiko/eksempler-
    pa-storulykke/deepwater-horizonmacondo/recommendationslessons-learned/chemical-
    exposure-in-connection-with-accidents-and-spills/. Accessed 09 Jan 2020
Parisi S (2012) Food packaging and food alterations. The User-Oriented Approach, Smithers Rapra
    Technology Ltd, Shawbury
Parisi S (2013) Food industry and packaging materials. User-oriented guidelines for users. Smithers
    Rapra Technology Ltd., Shawbury
Parisi S (2016) Prevenzione Incendi: Requisiti Legali e Approfondimenti nel Settore Chimico.
    Interprovincial Order of Sicilian Chemists, Palermo
Picot A, Grenouillet P (1994) Safety in the chemistry and biochemistry laboratory. Wiley, Hoboken,
    p 282
Puntarić D, Kos A, Smit Z, Zecić Z, Sega K, Beljo-Lucić R, Horvat D, Bosnir J (2005) Wood dust
    exposure in wood industry and forestry. Coll Antropol 29(1):207–211

Ross MJ, Godin C (2016) Guide entreposage des produits dangereux dans le secteur manufacturier. Multiprévention, Longueil. Available https://multiprevention.org/wp-content/uploads/2018/09/guide-multiprevention-entreposage-produits-dangereux.pdf. Accessed 10 Jan 2020

Safe Work Australia (2012) Managing risks of hazardous chemicals in the workplace—code of practice. safe work Australia, Canberra. Available https://www.safeworkaustralia.gov.au/system/files/documents/1702/managing_risks_of_hazardous_chemicals2.pdf. Accessed 08 Jan 2020

Silbergeld EK (1998) Toxicology. In: Stellman JM (ed) The ILO encyclopedia of occupational health and safety, 4th edn. International Labour Office, Genève. Available http://www.ilocis.org/documents/chpt33e.htm. Accessed 08 Jan 2020

SSTI (2013) Guide pratique d'évaluation et de prévention du risque chimique en enterprise. Fédération régionale des services de santé au travail (SSTI) des Pays de la Loire, Lava. Available https://www.risquechimiquepaysdelaloire.org/sites/default/files/guide_pratique_risque_chimique_ssti_pays_de_la_loire_-_edition_2013.pdf. Accessed 10 Jan 2020

Stellman JM (ed) (1998) The ILO encyclopedia of occupational health and safety, 4th edn. International Labour Office, Genève

United States Department of Labor (2013) Regulations (Standards—29 CFR), Part 1910—occupational safety and health standards. United States Department of Labor, Washington, DC. Available https://www.osha.gov/pls/oshaweb/owadisp.show_document?p_id=10099&p_table=STANDARDS. Accessed 08 Jan 2020

WHO (2016) Ambient air pollution: a global assessment of exposure and burden of disease. World Health Organization (WHO), Geneva. Available https://www.who.int/phe/publications/air-pollution-global-assessment/en/. Accessed 08 Jan 2016

WHO (2020) Endocrine disrupting chemicals (EDCs). World Health Organization, Genéve. Available https://www.who.int/ceh/publications/endocrine/en/. Accessed 09 Jan 2020

Ziem GE, Castleman BI (1989) Threshold limit values: historical perspectives and current practice. J Occup Med 31(11):910–918. https://doi.org/10.1097/00043764-198911000-00014

Chapter 2
Uses of Chemicals in the Food and Beverage Industry

Abstract Food additives, sanitizers, cleaning agents, technological aids, components of packaging materials and objects, and other definitions can be applied to chemicals when used in the industry of edible commodities—all edible products—for human and animal consumption. Nowadays, approximately 100% of food productions worldwide are realized totally or partially with the help of chemical sub stances. By the technological viewpoint, chemicals are generally needed because of two reasons at least (when speaking of foods and beverages): the reconstitution, treatment, or modification of one or more of involved raw materials because of different reasons, and an exigency of ameliorated foods with enhanced durability and/or new features (improved colors, aromas, excellent cooking resistance, etc.). Consequently, the modern food and beverage industry cannot be conceived today without chemicals: food additives, cleaning, and sanitizing agents, …, and a careful evaluation has to be carried out when speaking of work safety. This chapter is focused on the use of several chemicals in the food industry depending on their use and peculiar dangers for workers and the environment.

Keywords Chemical risk · Cleaning agent · European union · Food additive · Sanitizer · Technological aid · Toxicology

Abbreviations

Citric acid (INS code)	E330
International Numbering System	INS
Ultra high temperature	UHT

2.1 Food Production and Chemicals: A Strong Relationship

Food additives, sanitizers, cleaning agents, technological aids, components of packaging materials and objects, and other definitions can be applied to chemicals when used in the industry of edible commodities—all edible products—for human and animal consumption. In the last 100 years, chemistry has entered the world of foods and beverages with a clear result: Nowadays, approximately 100% of food productions worldwide are realized totally or partially with the help of chemical substances.

The same definition of 'Food Chemistry' means implicitly that food technological aspects are—or should be—studied with some relation to chemical transformations on the one side, and the concomitant use of ingredients which could not be defined exactly 'foods' or 'beverages. The origin of similar substances or mixtures (Chap. 1) is generally linked with the chemical industry, but with several differences depending on initial raw materials. According to common food consumers' belief, all components of foods and beverages should be 'natural' as synonym of materials with vegetable, animal, or mineral origin. In spite of this obvious opinion, the production of 'natural' ingredients involves the use of chemical, biochemical, and/or synthetic processes which may be questionable if seen with the eye of a common food consumer …

By the technological viewpoint, chemicals are generally needed because of two reasons at least (when speaking of foods and beverages):

(a) One or more of involved raw materials need to be reconstituted, treated, or modified chemically because of different reasons, differently from a recognized, historical or 'traditional' food production process.

(b) The exigency of new foods with enhanced shelf-life periods (enhanced durability) and/or new features such as improved colors, aromas, excellent cooking resistance, and diminished amount of tolerated sugars (such as low-lactose products) may force producers to use food-grade additives with the aim of obtaining desired results (by food consumers or food distributors).

Consequently, the modern food and beverage industry cannot be conceived today without chemicals. On the other hand, the importance of chemicals has to be considered when speaking of allowed uses because food additives—a useful example—are only a relevant portion of chemical substances and mixtures for the food industry…

The following sections are dedicated to a brief discrimination of chemicals for food-production-related purposes on the basis of their use instead of related dangers (Chaps. 1 and 3).

In general, the classification of similar chemicals in this (broad ambit) should follow a five-level structure (Laganà et al. 2019a, b, c, d; Ross and Godin 2016):

(1) Food additives (including technological aids—substances, mixtures, etc.) are not normally considered foods, and for this reason are not used as traditional ingredients, but needed for technological reasons. Examples could also concern smoking agents such as wood for burning operations.

(2) Food packaging components—substances needed for the production of contain-
    ers, objects, and materials for food-contact application. Naturally, these compo-
    nents should be part of packaging materials and objects instead of foods. How-
    ever, a little category of 'food additives' might be in relation with packaging
    technologies themselves. The so-called food gases for shelf-life enhancement
    (and for other reasons) are part of the packaged food, but their reaction with
    foods, packaging materials may be questionable… In relation to food operators,
    the application of gases such as argon, carbon dioxide, and nitrogen has to be
    considered with extreme care. On the other hand, these 'additives' are generally
    associated with packaging technologies (Laganà et al. 2019a, b, c, d).
(3) Cleaning agents—substances and/or mixtures—needed for the cleaning of
    systems, machines, and equipment used for food manufacturing, processing,
    preparation, treatment, packaging, transport, storage …
(4) Sanitization agents—substances and/or mixtures—needed for the disinfection
    of cleaned systems, machines, and equipment used for food manufacturing,
    processing, preparation, treatment, packaging, transport, storage… In addition,
    all systems used for water disinfection need to use chemicals such chlorine and
    other oxidant chemicals.
(5) Fuel and other flammable mixtures needed for the ignition and continuous
    function of thermal reactors, cooking machines, etc.

With relation to the aims of this book, only food additives, cleaning and sanitizing
agents should be considered, while:

(1) Food packaging components are completely associated with materials and
    objects for food-contact application. Food gases should be considered as 'food
    additives'.
(2) Fuels and other flammable mixtures are not related to foods and beverages
    themselves.

2.2   Food Additives

With concern to 'food additives,' it has to be considered that their use may occur
in many steps of food production: initial manufacturing, processing, preparation,
treatment, packaging, transport, storage … In addition, the presence of chemical
substances or mixtures may be reasonably correlated with the final product itself, so
that the analytical detection in relevant amount may be not surprising (and generally,
it is expected!). Finally, it has to be noted that the association of similar additives
with possible teratogenic, mutagenic, and carcinogenic risks is not allowed on the
regulatory level. Consequently, food additives should not pose risks (to food consumers, provided that regulatory limits are fixed and respected where needed), while
the use by food operators may reasonably cause problems, and this aspect has to
be clarified. The classification of these substances and mixtures may be simplified
enough taking into account the following groups (Laganà et al. 2017a, b, c, d):

(a) Food protection (preservatives, antimicrobials, antioxidants, coating agents, gases for modified atmosphere treatments, etc.)
(b) Colorants
(c) Sweetening agents
(d) Structure and technology enhancers (gelling agents, melting salts, propellants, stabilizers, thickeners, etc.).

By the chemical point of view, the property of each food additive is related to the declared use. Above all, the risk for food operators has to be linked with the use in a peculiar production step and with the physical form (liquid, solid, powder). By the technological viewpoint, the classification of these substances with the 'International Numbering System (INS) for Food Additives' is relevant enough (Laganà et al. 2017a, b, c, d) because of the link between INS numbers and their chemical functions with associated chemical attributions. On these bases, food additives may be roughly classified as follows (several categories may be not shown here):

(1) Acidity regulators: salts such as carbonates
(2) Carbon dioxide (CO_2) and carbonated generators (carbon dioxide itself, some carbonates)
(3) Salts with anticaking properties
(4) Waxes and other synthetic materials with antifoaming features
(5) Antioxidant agents such as strong acids (phosphoric acid, some salt)
(6) Oxidant agents such as chlorine
(7) Carriers, colorants (e.g., calcium lactobionate, potassium alginate)
(8) Emulsifying agents, flavoring agents, er (e.g., powdered cellulose)
(9) Foamers (nitrogen, an interesting 'food gas')
(10) Gelifying agents (several gums)
(11) General packaging gases or 'food gases,' including propellants (e.g., CO_2)
(12) Preservatives (some organic acids and salts are present here)
(13) Raising agents (several acids and salts).

This classification is extremely difficult also because one single chemical can be found in more than one single category. With reference to our aims, a specific acid with different functions (and some important safety norms when speaking of food operators) is discussed in Sect. 2.2.1.

2.2.1 Citric Acid

According to the 'Codex General Standard for Food Additives' (Alimentarius 1995), last revision 2018, citric acid (INS code: E330) can be used in the preparation of foods and beverages (selected categories and uses) with the following functions (which have to be declared on food labels):

(a) Acidity regulator
(b) Antioxidant

(c) Color
(d) Retention agent
(e) Sequestrant.

This chemical, when used as food additive, may be allowed in the following situations (the list is not exhaustive) (Alimentarius 1995):

(1) Milk protein
(2) Cheeses
(3) Fermented milks
(4) Plain pasteurized, sterilized, and/or ultra-high-temperature (UHT) creams
(5) Butter oil, anhydrous milk fat, *ghee*
(6) Vegetable oils, vegetable fats
(7) Lard, tallow, fish oil, and other animal fats
(8) Untreated fresh vegetables, seaweeds, nuts, and seeds
(9) Frozen vegetables, seaweeds, nuts, and seeds
(10) Fermented vegetables
(11) Fresh and dried pastas and noodles
(12) Fresh comminuted meat, poultry, etc.
(13) Fresh mollusks, crustaceans, and echinoderms
(14) Smoked, dried, fermented, and/or salted fish and fish
(15) Liquid or frozen egg products
(16) Salt substitutes
(17) Infant and follow-up formulae
(18) Fruit juices, vegetable juices (and related concentrates for similar preparations)
(19) Fruit and vegetable nectars (and related concentrates for similar preparations)
(20) Coffee, coffee substitutes, tea, herbal infusions, etc. (with the exclusion of cocoa).

Should citric acid be used in a preparation, its function would be declared and on condition that it is allowed with possible notes (e.g., maximum allowed amount, or *quantum satis* condition limited by Good Manufacturing Practices). Anyway, the classification of allowed used (for the production of selected products) does not impose certain functions. As an example, E330 may be used as acidity regulator in a cheese production and be used as sequestrant (in combination with other salts, possibly sodium citrates) in other (processed) cheese formulations (Mania et al. 2017).

On the other side, E330 has different intrinsic dangers for food operators. Consequently:

(a) Citric acid is allowed and safe in food productions for food consumers provided that certain precautions are observed, and
(b) At the same time, citric acid has to be managed safely by food operators with dedicated precautions, depending on the physical state of E330 and the interested production step (e.g., use of high temperatures? Possibility of aerosolized suspensions? Possibility of recirculating air? Presence of water? Presence of alkaline media in the immediate proximity?).

Allowed functions of E330 (citric acid monohydrate) in food products and correlated dangers for food operators should be considered. A simple analysis (derived from normal safety data sheets available for E330) demonstrates that dangers depend on several factors, but the targeted people under risk appear to be the main factor of the problem.

In brief (phrases have been modified with the aim of a simple comprehension) (European Parliament and Council 2008; SSTI 2013):

- Citric acid is classified as 'Serious eye damage/eye irritation, Hazard Category 2; hazard statement: H319' according to the Regulation (EC) No 1272/2008
- The following precautionary statements have to be considered:
  P280: Wear eye protection/face protection
  P305+P351+P338: Rinse cautiously with water for several minutes. Remove contact lenses, if present and easy to do. Continue rinsing (if E330 is in contact with eyes)
  P337+P313: Get medical advice/attention (if eye irritation persists).
  In addition, the following adverse reactions should be expected in order to implement adequate first aid measures:
- If consumption has occurred, rinse mouth with water (if swallowed); search for immediate medical help/advice and show related E330 container. Do not induce vomiting.
- If eye contact has occurred, first, rinse with plenty of water for several minutes and then search for immediate medical help.
- If inhalation has occurred, remove the person to fresh air and keep at rest.
- If skin contact has occurred, remove contaminated clothes; rinse skin with plenty of water or shower; then search for immediate medical help.

As a consequence (handling and storage precautions), it is recommended to keep away E330 from food, drink, and animal feeding stuffs, and do not smoke. Handling in accordance with good industrial hygiene and safety practice is extremely recommended. Also, E330 has to be and remain well sealed in its original container well sealed, in a dry and ventilated area, away from potential fire/ignition sources.

Some additional information can be provided when speaking of incompatible materials (oxidizing mineral acids, strong reducing agents, strong oxidizing agents) and hazardous decomposition products (CO_2 and other unidentified organic compounds which may be originated upon combustion).

2.3   Cleaning and Sanitizing Agents

In spite of their different action and effective chemical differences, cleaning agents—e.g., alkaline foaming agent with high sequestrant attitudes, good for cleaning-in-place (CIP) systems—and sanitizing agents—e.g., surface disinfectant and foam cleaner for manual or semiautomatic processes—may be confused. Actually:

(a) Cleaning agents are applied on surfaces with the aim of removing dirty particles and food-related materials by means of the action of both caustic alkaline media such as sodium hydroxide, and one or more sequestrant molecules
such as ethylenediaminetetraacetic acid (EDTA) sodium salt. This application is
assumed to be made after manual removal, and cleaning agent can have peculiar
dangers for food operators.

(b) Sanitizers are applied on surfaces generally after foaming agents and subsequent water washing, with two different aims: (i) to remove residual foam traces
and (ii) to sanitize permanently (for a certain number of hours, until the next
production cycle) surfaces. This operation is possible by means of the action
of quaternary ammonium salts, substituted diamines, polybiguanides, hydro-
genated betaines, etc. After this step, a final washing with water is required.
Sanitizing agents can have peculiar dangers for food operators.

The main dangers (in form of hazards and precautionary statements) of certain cleaners and sanitizers should be preventively considered. Actually, there are
a plethora of possible combinations and products on the market. Consequently, the
figure has to be intended only as an example of possibilities offered by both cleaning
and sanitizing agents.

A possible cleaning agent—actually, a mixture of non-dangerous and dangerous
substances, so it is a 'mixture'—can exhibit a safety data sheet (SDS) showing the
following classifications (among others):

- 'Causes severe skin burns and eye damage' 1 A (H314)
- 'May be corrosive to metals' 1 (H290)

and consequently recommend the following precautionary statements (among others,
phrases have been modified with the aim of a simple comprehension):

- P280: Wear eye protection/face protection
- P303+P361+P353: Take off immediately all contaminated clothing, then rinse skin
  with water or shower (if skin or hair have been in contact with the product)
- P305+P351+P338: Rinse cautiously with water for several minutes. Remove contact lenses, if present and easy to do. Continue rinsing (if the product is in contact
  with eyes)
- P310: Immediately search for a poison center and/or medical help.

At the same time, sanitizing agents may exhibit very similar data in their SDS,
with the possible addition (because of sanitizing active principles) of certain hazard
statements related to ecotoxicological considerations. One simple example is:

'Harmful to aquatic life with long lasting effects' 3 (H412), and this statement
can be added to above-mentioned phrases. For the rest, there is a certain similarity
between SDS of cleaners and sanitizers not because of similar features and composition, but because of the similar physical state and the premise that sanitizers have
to remove efficacy cleaners (after a preventive aqueous washing). Consequently, the
physical similarity of water-diluted solutions (rheology, etc.) is recommended and
also desired.

References

Alimentarius C (1995) General standard for food additives, CODEX STAN 192-1995. Codex Alimentarius Commission, last revision 2018. The Food and Agriculture Organization of the United Nations (FAO), Rome; The World Health Organization (WHO), Geneva. Available http://www.fao.org/gsfaonline/docs/CXS_192e.pdf. Accessed 08 Jan 2020

Laganà P, Avventuroso E, Romano G, Gioffré ME, Patanè P, Parisi S, Delia S (2017a) Classification and technological purposes of food additives: the European point of view. In: Chemistry and hygiene of food additives. Springer, Cham

Laganà P, Avventuroso E, Romano G, Gioffré ME, Patanè P, Parisi S, Delia S (2017b) The Codex Alimentarius and the European legislation on food additives. In: Chemistry and hygiene of food additives. Springer, Cham

Laganà P, Avventuroso E, Romano G, Gioffré ME, Patanè P, Parisi S, Delia S (2017c) Food additives and effects on the microbial ecology in Yoghurts. In: Chemistry and hygiene of food additives. Springer, Cham

Laganà P, Avventuroso E, Romano G, Gioffré ME, Patanè P, Parisi S, Delia S (2017d) Use and overuse of food additives in edible products: health consequences for consumers. In: Chemistry and hygiene of food additives. Springer, Cham

Laganà P, Campanella C, Patanè P, Cava MA, Parisi S, Gambuzza ME, Delia S, Coniglio MA (2019a) Food gases: classification and allowed uses. In: chemistry and hygiene of food gases. Springer, Cham

Laganà P, Campanella C, Patanè P, Cava MA, Parisi S, Gambuzza ME, Delia S, Coniglio MA (2019b) Food gases in the European union: the legislation. In: Chemistry and hygiene of food gases. Springer, Cham

Laganà P, Campanella C, Patanè P, Cava MA, Parisi S, Gambuzza ME, Delia S, Coniglio MA (2019c) Food gases in the industry: chemical and physical features. In: Chemistry and hygiene of food gases. Springer, Cham

Laganà P, Campanella C, Patanè P, Cava MA, Parisi S, Gambuzza ME, Delia S, Coniglio MA (2019d) Safety evaluation and assessment of gases for food applications. In: Chemistry and hygiene of food gases. Springer, Cham

Mania I, Barone C, Pellerito A, Laganà P, Parisi S (2017) Traspa-renza e Valorizzazione delle Produzioni Alimentari. L'etichettatura e la Tracciabilità di Filiera come Strumenti di Tu-tela delle Produzioni Alimentari. Ind. Aliment 56(581):18–22

Ross MJ, Godin C (2016) Guide entreposage des produits dangereux dans le secteur manufacturier. Multiprévention, Longueil. Available https://multiprevention.org/wp-content/uploads/2018/09/guide-multiprevention-entreposage-produits-dangereux.pdf. Accessed 10 Jan 2020

Chapter 3
The Durable Management of Chemicals in the Food Industry and Annexed Laboratories

Abstract At present, the use of chemicals in the food production environment is tacitly assumed when speaking of technological and hedonic advantages. On the other side, industrial foods and beverages are often questioned because of the possible presence of undesired chemicals without relation to the intended food or beverage (pesticides, bisphenol A, etc.). However, the problem of chemical risk assessment does not concern food consumers only, but also food workers in each area of production plants, including chemical and microbiological laboratories carrying on quality controls. This chapter concerns the reliable, safe, and durable management of chemicals in the food production and packaging areas on the one side, and in annexed quality control laboratories, highlighting the following points: toxicological risks; the importance of labeling; technical and material safety data sheets; risk categories (fire, explosion, unpredicted chemical reactions, etc.); safe use of hazardous chemicals, prevention procedures, and emergency planning in laboratories and industrial areas.

Keywords Chemical risk · CLP Regulation · European Union · Safety data sheet · Cancerogen · Mutagen · Teratogen

Abbreviations

AISS	Association internationale de la sécurité sociale
AEV	Average exposure value
CO_2	Carbon dioxide
CMR	Carcinogenic, mutagenic, reprotoxic
CAS	Chemical Abstracts Service
CLP	Classification, labeling, and packaging of chemical products
CIP	Cleaning-in-place
DCO	Dangerous Chemical Agent
EU	European Union
FAO	Food and Agriculture Organization of the United Nations
GHS	General Harmonized System

HACCP   Hazard analysis and critical control points
HF   Hydrofluoric acid
H_2O_2 Hydrogen peroxide
INRS   Institut National de Recherche et de Securitè
IARC   International Agency for Research on Cancer
LD_{50} Lethal dose 50
LC_{50} Mean lethal concentration
OEL   Occupational exposure limit
OHP   Occupational health physician
O_2 Oxygen
PBT   Persistent, bioaccumulative, and environmentally toxic
PPE   Personal protective equipment
SDS   Safety data sheet
STEL   Short-term exposure limit
TLV   Threshold limit value
UNEP   United Nations Environment Programme
vPvB   Very persistent, very bioaccumulable
WHO   World Health Organization

3.1   Introduction to Chemicals as Dangerous and Manageable Products

Regulations on the classification, packaging, and labeling of hazardous chemicals are intended to protect workers, consumers, and the environment. Labeling is the first information provided to the user on the dangers and the precautions to be taken during their use. The classification of chemicals (substances and mixtures of substances) must be based on inherent dangers to health and physical integrity, including (Anonymous 2012; ILO 1993; OSHA 2016):

1. Toxic products (acute and chronic health effects)
2. Chemical and physical characteristics (flammable, explosive, oxidizing properties, etc.)
3. Corrosiveness and irritant features
4. Allergenic and sensitizing effects
5. Teratogenic and mutagenic effects
6. Effects on the reproductive system.

One or more hazard classes are ascribed to each substance or chemical mixture, identifying the different types of hazards (flammable, corrosive, acute toxicity…) and, for each class, a hazard category, which defines the level of danger severity. This categorization is based on criteria such as the physical state (gas, liquid, solid, aerosol), the flammability criteria (flash point, boiling temperature, lower and upper limits), and stability criteria (presence of unstable chemical bonds). The classification of substances and mixtures is based on categories that take into account the degree of danger and the specific nature of hazardous properties.

The European (EC) No 1272/2008 concerning the classification, labeling, and packaging of chemical products (also defined CLP Regulation) (European Parliament and Council 2008; SSTI 2013) provides 28 classes of danger in which substances and mixtures are distributed, according to their physical effects and their effects on health and the environment. Substances and mixtures which meet the criteria for physical, health, or environmental hazards, as set out in the CLP Regulation, are deemed to be dangerous. It should be remembered that CLP Regulation (European Parliament and Council 2008; SSTI 2013) is inspired by a recommendation of the United Nations, 'The General Harmonized System' (GHS). These criteria (UN 2019) may be different depending on old or new classification systems, as shown in Fig. 3.1.

In general, the label constitutes the first essential and concise information provided to the user on dangers and on precautions to be taken during use. The CLP Regulation (European Parliament and Council 2008) defines how substances, mixtures and mixtures must be classified, labeled, and packaged. However, in the workplace or at home, there may be danger labels that meet the European pre-existing system according to the repealed Directive 67/548/EEC (Council of the European Economic

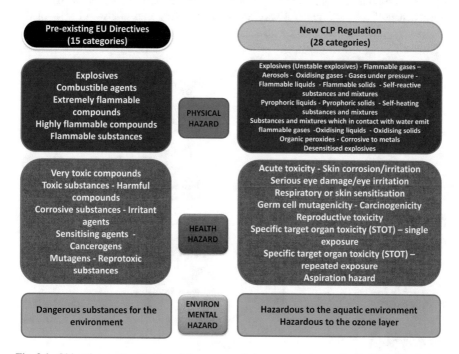

Fig. 3.1 Old and new classification of dangerous substances and mixtures in the European Union. This table has been realized by Carmelo Parisi, currently a student at the Liceo Scientifico Stanislao Cannizzaro, Palermo, Italy

Community 1967). Criteria are established to define a chemical membership in a class or category of hazard.

It should be noted that the classification and the definitions of these criteria applied by these systems vary in their number, degree of danger scales, terminology, test methods, and the method of classification of chemical mixtures. The setting of an international structure aiming at the harmonization of classification and labeling systems for chemicals would have a positive effect on trade in these products, the exchange of information about them, assessment costs, the management of associated risks, and the protection of workers, the general population, and the environment. As a result, the GHS classification and labeling need to be generalized and applied globally. It aims not only to ensure a high level of protection of human health and the environment through clear hazard communication, but also to facilitate trade in chemicals. For most chemicals, the classification is comparable, or in some cases, more severe than with the previous European Directive 67/548/EEC. Figure 3.2 shows old and new pictograms of the European Union for five dangers and associated chemicals (according to the new CLP Regulation).

This chapter is dedicated to the management of chemicals in food- and beverage-related industries, including also annexed analytical laboratories. The management of these compounds is the same of non-food enterprises. Consequently, the proposed approach will follow a general way, while considering that many—of all—of the discussed guidelines have to be also adapted for a food-related environment. This

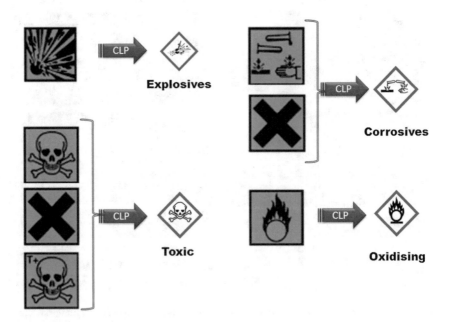

Fig. 3.2 Old and new pictograms for dangerous substances and preparations. Four examples

discussion means implicitly a 'food-centric' approach as recommended by national and international authorities in the ambit of food safety and public health (Parisi 2012, 2013; Parisi et al. 2016).

3.2 Information Documents

It is important to obtain all 'safety data sheets' corresponding to the hazardous products used in order to adopt an appropriate management mode (storage conditions, conditions of use, mode of intervention for incidents, etc.). The suppliers of chemical products must provide them request to recipients for information on hazards and properties (of the related chemicals), possible risks to human health and the environment, and protection measures and conditions of use (Ross and Godin 2016). These safety data sheets must be stored in a filing cabinet for reasons of traceability (a concept well known in food and beverage industries!) and must be present at each workstation where the concerned product is used (Anonymous 2012; ILO 1993; OSHA 2016; UN 2019).

Thus, the employer is required to determine whether (1) dangerous chemical agents are present in the workplace and assess (2) the risks they pose to the health of workers by taking into account (Brossat et al. 2012):

1. Their dangerous properties
2. Safety and health information provided by the supplier
3. Level, type, and duration of exposure
4. Conditions of use, including needed quantity, and this information is strictly required when speaking of quality management systems in the food ambit. Please take care that conditions of use do not contrast with formulations, where applicable!
5. The national occupational or biological exposure limit values (AISS 2016)
6. The effect of preventive measures
7. The conclusions to be drawn from a health surveillance that may have already been carried out.

3.3 The Safety Data Sheet (SDS)

Mandatory regulations (also named 'labor codes') require suppliers (importers or manufacturers) of chemicals to provide downstream users (including food producers, transporters, handlers, and storage facilities) with sufficient information to use chemical substances or preparations safely (Anonymous 2012; ILO 1993; OSHA 2016; UN 2019). The main tool for transferring information is the safety data sheet (SDS). These SDS must be sent to the occupational health physician (OHP) by the responsible of the industry.

The SDS is a multi-page document developed by manufacturers. It provides information on possible health dangers when using a chemical, how to protect yourself, and what to do in the event of an accident. It is an essential document for the prevention of chemical risk. The SDS is provided free-of-charge on paper or in electronic form. It is accessible at all stages, from the manufacture of the product to its disposal: manufacturing, distribution, transport, intermediate storage, use, and disposal. In most cases, it is provided on a mandatory basis, and in some cases, it will only be on request.

It should be noted that SDS must provide much more useful information on health, safety, and the environment than the label, which is primarily visual information. Complementary to labeling, the SDS is a valuable source of information for the evaluation of chemical risks and the various actors of prevention. SDS (European Commission 2013; SSTI 2013):

1. Recalls the effects on health (with pictograms and risk phrases of the labeling), allowing to know how the product is dangerous (carcinogenic effect? allergenic effect? health damage by inhalation or skin contact?)
2. Provides information on immediate prevention means and measures (ventilation? respiratory mask type? dilution? type of gloves such as nitrile or neoprene?)

In addition, SDS provides information on:

(1) The physical state of the product (very volatile liquid? powder? dough?…)
(2) The composition of the product.

Moreover, in the European ambit, SDS must (Anonymous 2012; European Commission 2013; ILO 1993; OSHA 2016; UN 2019):

(1) Be mandatorily transmitted with 'persistent, bioaccumulative, and environmentally toxic') (PBT) and 'very persistent, very bioaccumulable' (vPvB) substances or preparations containing such substances
(2) Be provided free-of-charge in the official languages of the European Union (EU) Member States in which the substance or preparation is placed on the EU market
(3) Be updated if new data on hazards or information that may affect risk management measures become available, if an authorization is granted or refused, or if a restriction has been put in place
(4) Contain in particular the following data:

    (4.1) Identification of the substance/preparation and of the company
    (4.2) Hazards identification
    (4.3) Composition information concerning the components
    (4.4) First aid to be carried in case of emergency
    (4.5) Fire-fighting measures, including even the prevention of explosions and fires
    (4.6) Accidental release measures
    (4.7) Precautions for use, handling, and storage
    (4.8) Exposure controls/personal protection

(4.9)   Physicochemical properties
(4.10) Product stability and reactivity
(4.11) Toxicological information
(4.12) Ecological information
(4.13) Disposal considerations
(4.14) Transport information
(4.15) Regulatory information
(4.16) Other information that may contribute to the general safety of the business and its environment.

Many of these sections provide usable information for identifying and quantifying hazards, while others contribute more to the choice of protection and response measures.

In addition, when exposure scenarios are developed as part of a chemical safety assessment, they should be attached to SDS and forwarded along the supply chain. In this ambit, the supplier informs the customer of implemented or recommended risk management measures with the aim of ensuring the safe use of the substance(s).

SDS must be updated (less than two years as a general EU rule) if:

(1) The official EU Regulation on the classification of substances is evolving (with scientific knowledge)
(2) The compositions can be modified
(3) The means of protection can be supplemented by new technical knowledge.

3.4   Classification, Packaging, and Labeling

The chemical label is an important source of information. Labels on containers of hazardous chemicals are the first warning that a chemical is hazardous; they should provide basic information on hazards of the substance, safe handling procedures, protective measures, and first aid in emergencies. They should also identify the hazardous chemical(s) and the name and address of their manufacturer. It represents the first, essential, and concise information provided to the final user on dangers and precautions to be considered when using a dangerous product.

By a general viewpoint, the labeling includes sentences as well as graphic symbols and color codes placed directly on the product container, packaging, or label. The marking should be easily understandable and able to withstand harsh climatic conditions (and this aspect has to be considered carefully when speaking of food- and beverage-related warehouses…). It should be placed on a background which contrasts with the color of the packaging or the card accompanying the product. SDS provides more detailed information on the nature of the risks associated with the product as well as the appropriate safety features (Anonymous 2012; ILO 1993; OSHA 2016; UN 2019).

With the new GHS-based regulation for the classification and labeling of chemicals (UN 2019), the information to be mentioned on the label changes. It contains, among others, the following elements (SSTI 2013):

(1) The regulatory label of a product. It may contain, in its security part, four elements, namely:

    (1.1) Danger symbols in the form of a danger pictogram. The pictogram represents each danger of the product

    (1.2) The signal word indicating the relative degree of danger, either 'DANGER' (for the most dangerous chemicals) or 'CAUTION'

    (1.3) The classification (based on dangers).[1]

In addition, the nature of particular risks attributed to hazardous substances and preparations is characterized (UN 2019) by hazard and precautionary statements. Hazard statements provide information on physical, health, and environmental hazards. These phrases indicate the severity of the hazards (e.g., moral, toxic, or harmful) and how they may occur (e.g., by route of exposure: by ingestion, skin contact, and inhalation). Precautionary statements indicate how to store, handle, dispose of chemicals, and describe first aid measures. These tips are essential for protecting health and the environment. They specify, among other things, how to use the product to minimize risks, for example, ventilating the premises or wearing gloves, clothing, glasses, and protective helmets.

---

[1] In the European Union, the following *modus operandi* was observed until 2007 (the CLP Regulation was issued in 2008):

E: Explosives. These substances explode under the effect of a flame or are very sensitive to shocks and friction, e.g., nitroglycerin

O: Oxidizing substances. Product able to generate fire without oxygen (O_2) input, e.g., hydrogen peroxide (H_2O_2) or potassium nitrate

F+: Extremely flammable compounds. Flash point <0 °C; boiling point <35 °C, e.g., molecular hydrogen, acetylene

F: Highly flammable compounds. Substances meeting one of the following criteria:

- Spontaneous heating and ignition at room temperature and without energy supply
- Solid, which can ignite by ignition and then continue to burn without ignition
- Liquid, having a flash point <21 °C
- Flammable gas at room temperature and atmospheric pressure
- Release of flammable gas with water (e.g., acetone)

Flammable agents (no pictogram), with 21 °C < flash point < 55 °C, with risk phrase 'R10: Flammable'

T^+: Very toxic compounds. Risk of serious intoxication, even fatal, by inhalation, ingestion or skin penetration even in very small quantities, e.g., hydrocyanic acid

T: Toxic substance. The same thing for: same as T^+ but for small quantities, e.g., methanol, ammonia

X_n: Harmful compounds. Risk of intoxication by inhalation, in-management, or skin penetration, e.g., iodine, xylene, tetrachloroethylene

X_i: Irritant agents. Risk of inflammatory reaction on contact with the skin, mucous membranes, e.g., bromobenzene

C: Corrosive substance. Risk of destructive action on living tissue

N: Dangerous substance for the environment. Risk of pollution of the environment.

Anyway, the symbol is added to the pictogram with the aim of distinguishing hazards or levels with the same pictogram (SSTI 2013).

At present, the new European Regulation known as 'CLP' from December 2010 for substances and June 2015 for mixtures has been implemented (European Parliament and Council 2008; SSTI 2013). For this reason, we will no longer speak of 'preparation' but of 'mixture'.

The new classification makes it possible to determine the different classes or categories of danger. Each danger category is associated with symbols and indications of danger and hazard statements. These elements constitute the classification of a substance or preparation. They should appear on the regulatory label with precautionary statements, originally named S-phrases in the EU and chosen on the basis of no longer in use risk phrases (R-phrases: presented in Fig. 3.2, black and yellow colors). In detail, the new categorization introduces:

(1) New terminologies ('mixtures' instead of 'preparations'; the term 'hazard category' is replaced by 'hazard class', etc.)
(2) New definitions of danger (Fig. 3.1). The 15 current danger categories are replaced by 28 danger classes, 27 of which are defined by the GHS and an additional class specific to the European Union entitled 'dangerous for the ozone' (UN 2019)
(3) New classification criteria
(4) New labels (danger pictograms, risk and danger phrases, precautionary statements), as shown in Fig. 3.1 (red, white, and black symbols). The danger symbol is always placed inside a square placed on the point, making it a diamond. The symbols must be black, and the diamond is represented by a red frame.

Interestingly, and with specific relation to food industries, chemical additives have a double-way label showing both safety information and food-related info, such as simple ingredients: composition, possible claims, lot, expiration date, etc. These labels might be really challenging, demonstrating that a specific training has to be supplied to food operators.

3.5 General Provisions Concerning Labeling

In the EU, the competent national authority of a Member State (or a body approved or recognized by the same authority) has to establish requirements for the marking and labeling of chemicals with the aim of allowing safe and secure use of chemicals. Chemical suppliers should ensure that chemicals are marked and hazardous chemicals are labeled, and that revised labels are prepared and provided to employers whenever new safety and health relevant information are available (European Parliament and Council 2008).

Should users receive chemicals that have not been labeled or marked, they should not use them until they have obtained the relevant information from the supplier or another reasonably accessible source. The information should be obtained mainly from the supplier. However, it can also be obtained from other sources (information sources and databases), so that before any use, labeling and marking can be carried out, in accordance with the requirements of the competent authority.

Naturally, these information do not concern nutritional data, lot identification, expiration dates, and other data which are normally supplied to food operators/producers in the ambit of public hygiene, food safety, and quality management. This clarification is important because many chemicals currently used in the food and beverage industry are used as food-grade additives or technological aids: In this situation, a double-way (work safety and food safety information) label should be provided.

3.6   Nature and Type of Allowed Labeling

All chemicals should be marked to allow their identification. The chosen marking should allow users to distinguish between chemicals when they are received, handled, and used. The marking should include the chemical identity, the common name, the registered trademark, the name or the code number, or any other designation, provided that the identity thus established is unique and, in the case of a dangerous chemical, the same than that which appears on the label and the safety data sheet. The mention of the name of the supplier on the transported container or on the packaging is recommended. As above mentioned, food additives should contain also information related to their food-grade status, such as traceability data (Mania et al. 2017; Pellerito et al. 2019a, b, c, d).

The labeling (marking) of chemical waste should indicate that it is waste (OSHA 2016). Labeling of chemicals may be impossible due to the size of the container or the nature of the packaging. However, these products should be easily identifiable by other means such as mobile labels or package inserts. Interestingly, this marking type is recommended and 'mandatory' (in the limited ambit of voluntary quality certification systems) when speaking of food industries: All possible waste materials and non-reusable intermediates have to be classified as 'waste.' Otherwise, the risk of contamination and/or possible reuse in the initial production cycle may be notable.

Each container (or each layer of the packaging) should be marked, as shown in Fig. 3.3. The indications given should be visible always on the container or packaging at least until the final use.

Dangerous chemicals should be labeled, in accordance with national laws and practice procedures, in such a way as to provide them with essential information and to report their identity in a way that is easily understood by workers who have to use them. However, there are possible differences when speaking of transport of chemicals which the information on the container or packaging may be different for.

The label should give essential information (ILO 1993; OSHA 2016):

1. On the classification of the chemical (and on the claimed food-grade status, where applicable)

Fig. 3.3 Each container (or each layer of the packaging) concerning chemicals should be adequately marked. The mandatory indications given should be visible always on the container or packaging at least until the final use, such as in this picture showing a general barrel

2. On its own dangers
3. On needed precautions.

In addition, this information should indicate the hazards of acute exposure and chronic exposure.

Labeling requirements, which should be in accordance with national requirements, should also include:

(1) In relation to required information, and where appropriate:

    (1.1) Trade names
    (1.2) The identification of the chemical
    (1.3) The name, address, and telephone number of the supplier
    (1.4) Danger symbols
    (1.5) The nature of particular risk(s) associated with the use of the chemical product
    (1.6) Safety precautions
    (1.7) The identification of the lot and expiration date
    (1.8) An indication that a safety data sheet providing additional information is available
    (1.9) The classification assigned in accordance with the system established by the competent authority

(2) With concern to readability, durability, and size of the label, appropriate information
(3) With concern to consistency of labels and symbols, needed data have to include used colors (with reference to food-related information, the matter of colors is not applicable in general).

Peculiar information concerning work safety should include:

(1) The indication of the concentrations of solutions, isomers, and components of petroleum distillation products and reactive chemicals when these information may have an influence or be correlated with properties of the chemical. A food-related example is the recommended dilution for cleaning solutions and disinfection liquids when speaking of 'cleaning-in-place' (CIP) systems.

(2) The identification of any component present which is likely to contribute significantly to the properties of the mixture, or which exceeds the concentration limit approved or recognized by the competent authority, in the case of mixtures.

(3) Similar information should be given in accordance with national requirements and taking into account the United Nations (UN) recommendations on the transport of dangerous goods, when speaking of transportation. The information provided should not only inform the carrier about chemicals, but also provide easily understandable indications to emergency services in the event of an emergency, in which case these indications would also be useful.

(4) In the case of waste, the information provided should include the telephone number of any person able to provide further information on the probable composition of waste and on related risks (when complete labeling is not available).

With peculiar relation to the last point, hazardous components of a waste should also be indicated, as far as possible and whether they affect the properties of the waste or exceed concentration limit agreed or recognized by the competent authority. The container or packaging should be appropriately labeled.

Unfortunately, labeling of hazardous chemicals may be impossible due to the size of the container or the nature of the packaging. In these situations, mobile labels or accompanying instructions should always be used. In such circumstances, all containers of hazardous chemicals should at least bear the hazard of the contents by means of appropriate markings or symbols. All containers and all layers of packaging should be labeled. The information given should be visible at all times on the container or packaging at all stages of the delivery and use of the chemicals.

Labels of pesticide containers may include additional information in accordance with applicable international guidelines such as those of the Food and Agriculture Organization of the United Nations (FAO) on good labeling practices. This disposition is particularly important in relation to agricultural enterprises because of the increasing amount of used pesticides worldwide.

3.7 Handling of Chemicals and Safe Logistic Procedures

Employers should ensure that, when chemicals are transferred to other containers or equipment, workers are correctly informed on the identification of these chemicals, and related risks/safety precautions, on the basis of the previous sections. When chemicals are transferred to other containers or equipment for use on the employer's premises, new containers or equipment should be also marked for identification (Brossat et al. 2012). In the case of hazardous chemicals, they should be labeled or carry any other indication allowing workers to identify:

(1) Chemicals, for example, from the reference number, code, or common name known to all workers elsewhere in the workplace
(2) Related dangers, for example, by means of appropriate inscriptions or symbols
(3) Safety precautions.

Interestingly, the 'common name' known to all workers may generate confusion. The normal experience in food-related enterprises may concern—as a simple and 'golden' example—the conviction of food operators that alkaline substances such as sodium hydroxide are named 'acids' and *vice versa*…

Moreover, a number of operations, installations, and equipment (e.g., reactors, distillation columns, and melting cookers in certain food facilities) may involve the treatment or handling of several different chemicals (including additives). When the marking or labeling of installations or equipment is impossible due to the variability of circumstances, workers should be informed of the identification of chemicals, inherent dangers, and precautions for security. In addition, a dedicated training in this area should be provided.

3.8 Chemicals and Related Labels

Hazardous chemicals must be labeled in accordance with regulatory obligations, country by country, with the aim of providing essential information and identifying these substances a way that is easily understood by users.

For example, labeling provides essential information on the classification of the chemical, its hazards, and the precautions to be taken. The chemical labels include pictograms, hazard statements which indicate the nature of the particular risks attributed to dangerous substances and preparations, and precautionary statements concerning dangerous substances and preparations.

Labeling provides also additional information on handling and hazards of chemical products. Danger symbols are a first level of information, of a purely visual nature, destined directly to the operator who seizes the packaging.[2]

It has to be noted that the absence of a danger symbol does not mean that the product is safe: in fact, safety regulations only require the symbol and the name of the dangerous substance to be indicated above a certain concentration, a certain degree of flammability, or a certain threshold of harmfulness. Any substance for which there is no information relating to dangers must be considered a priori dangerous.

The label or the inscription must be affixed so as to be very visible and readable horizontally when the packaging is in normal position.

---

[2]A useful guideline concerning the complete information for CLP-style labels can be found at the following web address: https://www.hsa.ie/eng/Publications_and_Forms/Publications/Chemical_and_Hazardous_Substances/CLP_info_sheet.pdf.

3.9   Storage and Compatibility of Chemical Products

Some products may react violently with each other, so they should not be stored in the same place (Fig. 3.4). Basically, this precaution is well known in chemical laboratories, but food-related environments are likely 'crowded' enough when speaking of food operators and stored (or moving…) commodities. Consequently, this recommendation should be repeated *ad libitum*…

When a product has more than one of risk classes previously defined, hazardous products have to be stored carefully (Ross and Godin 2016). The best strategy is the separation of incompatible products in order to limit the impact of an accidental spill and to reduce the risk of fires or violent reactions. A malfunction among these incompatible chemicals can cause them to come into contact. They can react with each other, sometimes causing explosions, fires, spatter, or dangerous gas emissions. Therefore, incompatible products should never be stored side by side but should be physically separated (Fig. 3.4). An example of avoidable accidents is the 'normal' reaction between an acid (citric acid…) and an alkali agent (sodium hydroxide) into storage areas because of sudden reaction (the real cause is generally the excessive closeness!).

Incompatibilities between chemicals have to be identified! Generally, safety data sheets accompanying any dangerous chemical are used for this preliminary evaluation (searching for specific incompatibilities to the product or chemical family). For more security, the storage room should meet the following criteria (ILO 1993):

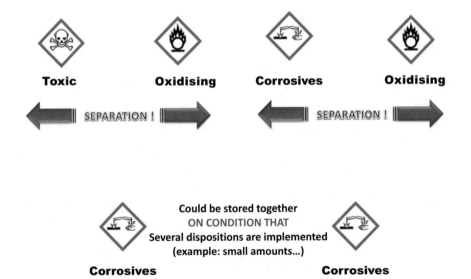

Fig. 3.4 Incompatibility of chemicals. Three examples can be displayed here showing that certain chemicals must be physically separated when speaking of storage, or they must not be stored together unless certain special provisions are applied (e.g., small quantities)

(1) It can be locked
(2) The ventilation must be adapted depending on circumstances.

Some recommendations can be displayed here:

(1) High and low natural ventilation is sufficient if needed.
(2) The room must be flameproof and have mechanical ventilation if one or more of stored chemicals belong to old E, F$^+$, or O categories. However, if the quantities are low, products may be stored in dedicated cupboards. Anyway, holding tanks have to be available and ventilated.
(3) Liquid products must be placed on a holding tank with volumetric capacity corresponding to at least stored volumes.
(4) A fire extinguisher must be located near the room as well as a water point. Shelves must be metallic. Absorbent materials must be present in the room. In addition, smoking will not be allowed in the room.

Moreover:

(1) Storage must separate chemical bottles from other utensils: rags, brushes, sponges, etc.
(2) A periodic removal of expired products, unused or no-longer-needed chemicals, containers without label (or damaged/unreadable labels), containers of residues and hazardous waste accumulated over time.

An important recommendation concerns phytosanitary products. These substances must be stored separately from other products in a specific and ventilated room and be kept in their original packaging until they are used (packaging used for handling purposes must have the same guarantees as original packaging). In this room, utensils reserved for the use of products have to be always available.

3.10 Chemical Safety

First of all, it should be noted that there is no harmless chemical. The nuisance potential of a chemical is a function of the type of risk presented by it (health, environment, safety), and it depends on stored and used quantities, frequency of use, etc. As a result, staff exposed to this risk factor must benefit from special medical supervision and appropriate training.

Chemical safety is based on two essential principles (ILO 1993; OSHA 2016; UN 2019):

(1) Limitation of exposure to harmful chemicals for employees and the environment using appropriate equipment and methods
(2) Knowledge of actions, containment measures, and best protection practices during accidental exposition to these chemicals.

Chemical risk management requires that:

(1) The hazards of the products must be clearly known and notified to all users: up-to-date SDS in the language of the country and labeling according to existing regulations.

(2) Personal protective equipment (PPE) must be worn and available. The staff must be sensitized (trained) and must observe protection rules always.

(3) A single document gathering all information related to risks, in particular the chemical risk with all the necessary documentation (safety data sheets or SDS, exposure sheets, follow-up of personnel, driving in the event of an accident, etc.), has to be available.

(4) Essential information and equipment adapted to products handled in the event of an accident must be promptly available.

These few rigorous rules are the basis for good control of chemical risk.

3.11   Chemicals and Safe Procedures. Dedicated Guidelines

In relation to the use of chemical products in all possible environments (academia, industry, laboratories, food companies, etc.), it is essential to adhere to a series of basic principles when using chemical products, especially when carrying out hygiene and sanitation tasks. A selection of guidelines workers should always observe in a general ambit (ILO 1993; OSHA 2016; UN 2019), and in the food and beverage ambit in particular is available in the next sections (because food operators use many different chemicals as food additives, cleaning systems, disinfection agents, … and many, many liters of water, the most common solvent for all chemicals, with possible displacement of dangers everywhere!).

3.11.1   Simplified Principles

(1) Know dangers and precautions related to the use of products.

(2) Read product labels, material safety data sheets and use only labeled containers, including bottles with spouts.

(3) Wear—when required—protective equipment (e.g., gloves, mask, and blouse).

(4) Always use chemicals only for the purposes which they are manufactured for.

(5) Refer to surface manufacturer's recommendations when speaking of amounts needed to maintain floor (in general for cleaning/sanitizing/disinfection operations) with the aim of avoiding damaged surfaces.

(6) Do not mix chemicals unless there is a recommendation or guideline by the manufacturer.

(7) With chemicals, use warm water to reduce the release of toxic vapors and maintain the effectiveness of the products.

(8) Avoid splashing when preparing solutions, and always pour a chemical into the water and not vice versa; any splashes will then contain a more dilute solution. Interestingly, this 'chemical' rule should be repeated always in spite of its apparent simplicity.

(9) Strictly observe instructions relating to dilution and needed quantities when speaking of pre-impregnation operations on production lines.

(10) When cleaning a surface with a cloth, impregnate the cloth instead of the surface to be cleaned in order to avoid any emanation.

(11) Limit spraying only to the maintenance of inaccessible spaces with a cloth; ensure that the sprayed product is used safely.

(12) Check the useful life of a product after dilution to ensure the effectiveness of the solution used; refer to manufacturer's information.

(13) Always respect the contact time of used products. Germicides should never be wiped off.

(14) Store products in a secure and safe way.

3.11.2   The Sinner's Circle

A perfect cleaning requires the use of products fit for the desired action(s): degreasing, scrubbing, descaling, etc. Moreover, the type(s) of cleaning tools have to be considered. To accomplish this mission, the 'Sinner's Circle' considers four simple factors which, when used together, will provide an undeniably more efficient cleaning job than any detergent used alone (Von Rybinski 2007). These four inseparable elements are:

(1) Chemistry (chemical action)
(2) Mechanical power (action)
(3) Contact time
(4) Temperature.

Under certain conditions, the increase of one of the parameters can improve the result of the whole.

3.11.2.1   Chemical Action

Any product is categorized according to its chemical action (e.g., detergent, floor finish, and disinfectant). Each chemical has a specific use that must be respected in order to maximize its effectiveness and avoid any accident.

Each product has a given efficiency according to a specific dilution; it is important to respect the dilution of each product, as overdose and underdosing may affect the expected result.

The same chemical may have different usefulness depending on the chosen dilution; it is therefore important to read the technical sheet. Moreover, chemical properties of water (e.g., its alkalinity or hardness) can significantly impact on the chemical effectiveness of a product.

3.11.2.2   Mechanical Action

The mechanical action on a surface can be expressed in different ways. Manual equipment (e.g., brushes and mops) is used to scrub surfaces on which the use of a machine is not applicable. Electromechanical equipment such as the low-speed polisher or the car washer generates greater pressure and friction.

During cleaning operations, the mechanical action must be modulated in order to avoid surface alterations. Actually, and with specific relation to food-related machines and equipment, the main problem is the damage, e.g., partial demolition of Teflon-coated surfaces for melting machines (cheese-making industries); fractures of plastic molds (various equipment and food-related productions) after repeated cleaning with mechanical tools; etc.

3.11.2.3   Contact Time

Contact time defines the required time of presence of a chemical on a surface for optimal efficiency. During cleaning and disinfection operations, the respect of contact times ensures the desired result of the chemical action.

On the other side, failure to respect the contact time can prevent the chemical from acting properly: Little or no effect if the product is not left long enough or possibility of damaging the surface if the product is left too long time.

Some chemicals remain 'effective' once dried (e.g., quaternary ammonium salts), making it easier to reach their contact time. On the other hand, other products must remain in aqueous solution to reach the necessary contact time for their full effectiveness (e.g., chlorine).

3.11.2.4   Temperature

Thermal action influences the qualitative result of cleaning activities. The thermal action can be generated by the temperature of the solution or by a mechanical action such as the friction of a pad on a ground surface. Some chemicals are effective within a certain temperature range.

The thermal action favors the action of thermoreactive products such as the products used during the dry-scrubbing technique of floor surfaces.

3.11.3 Dilution

Dilution allows to reduce the concentration of a solution. As a basic principle of dilution, any chemical should be diluted according to the manufacturer's recommendations with the aim of reaching the maximum effectiveness.

There are two ways to dilute a chemical (Kenkel 1992; Stockwell 1996):

(1) Automated dilution. In this way, the use of a diluter is needed. Because of the ameliorated control of the quality of dilution, this is the recommended strategy. If possible, use diluents to have sealed chemical containers (better conservation). It is important to ensure the effectiveness of the dilutor by performing solution tests periodically according to supplier's recommendations. The risk of splashing is very limited, representing a significant gain for the safety of workers and customers. Cost controls are easier when speaking of automatic dilution (if compared with manual dilution) through more efficient consumption of chemicals.

(2) Manual dilution. There is a possibility of error in the dilution because of the lack of precision (too diluted or too concentrated solutions). Either a graduated container, a pump, or any other measuring device must be used in this ambit. There is a significant risk of splashing the solution when pouring. The use of protective equipment (e.g., glasses, gloves, and mask) may be necessary.

3.11.4 Technical and Organizational Preventive Measures

For all activities possibly presenting a risk of exposure to dangerous chemical agents, the employer implements detailed safety measures as explained in the following lines (Cohrssen and Covello 1999; Crowl and Louvar 2001):

(1) Limitation of the number of exposed workers
(2) Measurement of workers' exposure, in particular with the aim of detecting abnormal exposures resulting from an accidental event
(3) Collection of pollutants and ventilate the workplace
(4) Implementation of appropriate work methods and procedures
(5) Implementation of collective protection measures, and if that is not enough, provide workers with personal protective equipment
(6) Ensuring basic hygiene
(7) Training for workers
(8) Delimitation and report actions concerning risk areas

(9)   Creation of emergency measures, especially in case of failure of closed systems
(10)   Use of airtight and labeled containers for storage
(11)   Safe handling and transportation
(12)   Creation of safe storage and disposal procedures in relation to waste.

Likewise, measures are taken to prevent risks linked to storage and handling of products, and the risk of fire/explosion. Workers carrying operations in confined spaces (certain food-production- and storage-related areas are good examples…) must be secured or protected by another safety device. The head of the external company responsible for the maintenance of PPE and work clothes is informed of the possible risks of contamination. Chemicals should be managed with care, especially CMR.

3.11.5   CMR Risk Prevention Strategy

The 'carcinogenic, mutagenic, reprotoxic' (CMR) risk prevention approach requires the systematic identification of all CMR agents present in the workplace. Risk prioritization is established taking into account the CMR classification, the duration of exposure, and the number of concerned people. The removal or substitution of a CMR agent by a less hazardous agent will be considered always.

The assessment of residual risks will make it possible to put in place appropriate prevention actions and the implementation of useful recommendations as follows (Anand 2011; Roy 2009; Ross and Godin 2016):

1.   Keep products out of reach of children (please note that this recommendation is not apparently of interest in our situation. However, the possible visit of many scholars in certain food-production-related areas has to be always considered… and similar precautions can be useful in these situations!).
2.   Keep (chemical) products in their original packaging.
3.   Choose products with solid, waterproof packaging, and convenient handling.
4.   Identify utensils used to prepare (chemical) products.
5.   Do not 'overcrowd' the shelves.
6.   Put clothing and personal protective equipment in a room.
7.   Classify (chemical) products according to their use (herbicides, fungicides, insecticides, toxic, flammable, harmful, irritants, explosives, etc.).
8.   Group on a shelf (in height) the most dangerous products on the higher place (especially toxic products).
9.   Never remove labels on packages.
10.   Arrange the different products, so that their labels can be read.
11.   Maintain accessibility, to avoid risky handling and writhing, common sources of spills or breakages.
12.   Do not smoke in the storage room.
13.   Maintain the electrical installation in good condition.

14. Display the reminder of the safety instructions in the room, as well as the emergency numbers (doctor, firefighters, poison control centers…).
15. Keep a log of stored products (dates of purchase, dates of use, quantity of products stored, etc.).
16. Train staff responsible for storage.

3.11.6 Further Recommendations

Beyond technical or organizational measures, a responsible company can also endeavor to promote a healthier lifestyle among its employees, to make them responsible for health matters, with particular emphasis on the prevention of serious and frequent illnesses and the benefits of a healthy lifestyle. The personal hygiene of employees contributes to hygiene at work. Working conditions have an impact on health, and food-related quality certification systems are remarkably based on hygiene recommendations.

Conversely, individual behaviors in terms of wearing personal protective equipment, hand washing, respiratory hygiene, hydration, nutrition, sleep or physical activity influence health and safety at work and must be adapted to the professional activity. Body hygiene is one of the means of prevention against occupational diseases.

Dirty hands are a vector of contamination, for oneself or one's entourage (private or professional), sometimes through everyday objects (door handles, tools, pencils, telephones, …). Work clothes can also be contaminated: They then represent a risk for both the employee and those around him.

Hygiene at work necessarily involves regular cleaning of hands, clothing used during work, and personal protective equipment. This measure is 'mandatory' in food and beverage companies in the ambit of hazard analysis and critical control points (HACCP) strategies.

From now on, bacteria, viruses, chemical or biological products, soiling, dust, etc., can carry a significant number of chemical or biological substances potentially dangerous for health by hands and work clothes. Some hygiene measures, helped and supported by an appropriate organization, can prevent some of the risks associated with these occupational nuisances. Occupational hygiene is based in particular on:

1. Individual behaviors (wearing adapted protective equipment, hand washing, nutrition, hydration and sleep adapted to the constraints of work, physical activity)
2. Adequate premises (sanitary facilities, ventilation and sanitation of air, break and catering premises, etc.)
3. Regular maintenance and cleaning of premises and work equipment.

Questions concerning personal hygiene are difficult to tackle because they fall within the personal sphere of employees. Companies can give their employees some advice (awareness, display, information, etc.). Therefore, the employer is required to implement measures to reduce occupational risks, to inform employees of pre-ferred behaviors, and to provide appropriate equipment (Brossat et al. 2012). Sanitary facilities include sinks, toilets, changing rooms, and—where appropriate—showers in sufficient number and in good working order. Organizational measures must be based on a relevant analysis of working conditions.

References

AISS (2016) Valeurs limites d'exposition pour la prévention des agents chimiques. Associa-tion internationale de la sécurité sociale (AISS), Institut National de Recherche et de Securitè (INRS), Paris. Available http://www.inrs.fr/dms/inrs/CataloguePapier/ED/TI-ED-6254/ed6254.pdf. Accessed 10 Jan 2020

Anand NP (2011) Hospital Sterilization. Jaypee Brothers Publishers, New Delhi

Anonymous (2012) Chemical and hazardous materials safety. Department of Environmental Health and Safety, The University of Texas at Dallas, Richardson, TX. Available https://www.utdallas.edu/ehs/download/Chemical_and_Hazardous_Materials_Safety.pdf. Accessed 09 Jan 2020

Brossat C, Thomasset L, Cote D, Certain E (2012) 10 bonnes Pratiques favorisant la santé au travail en contribuant à la performance globale des PME; Expériences conduites en Rhône-Alpes. SP1177—May 2012. Carsat Rhône-Alpes, Direction des Risques Professionnels et de la Santé au Travail, Lyon Cedex. Available https://www.carsat-ra.fr/images/pdf/entreprises/sp1177.pdf. Accessed 10 Jan 2020

Cohrssen JJ, Covello VT (1999) Risk analysis: a guide to principles and methods for analyzing health and environmental risks. United States Council on Environmental Quality, Washington, DC

Council of the European Economic Community (1967) Council Directive 67/548/EEC of 27 June 1967 on the approximation of laws, regulations and administrative provisions relating to the classification, packaging and labelling of dangerous substances. Off J Eur Comm No 196:1–98

Crowl DA, Louvar JF (2001) Chemical process safety: fundamentals with applications. Prentice Hall PTR, Upper Saddle River, NJ

European Commission (2013) Chemicals at work—A new labelling system. Guidance to help employers and workers to manage the transition to the new classification, labelling and pack-aging system. European Commission, Directorate-General for Employment, Social Affairs and Inclusion, Unit B3, Brussels. Available https://osha.europa.eu/it/file/49187/. Accessed 08 Jan 2020

European Parliament and Council (2008) Regulation (EC) No 1272/2008 of the European Parliament and of the Council of 16 December 2008 on classification, labelling and packaging of sub-stances and mixtures, amending and repealing Directives 67/548/EEC and 1999/45/EC, and amending Regulation (EC) No 1907/2006. Off. J Eur Union L 353:1–1355

ILO (1993) Safety in the use of chemicals at work: An ILO code of practice. International Labour Office, Geneva. Available https://www.ilo.org/wcmsp5/groups/public/—ed_protect/—protrav/—safework/documents/normativeinstrument/wcms_107823.pdf. Accessed 09 Jan 2020

Kenkel J (1992) Analytical chemistry refresher manual. CRC Press, Boca Raton

Mania I, Barone C, Pellerito A, Laganà P, Parisi S (2017) Trasparenza e Valorizzazione delle Produzioni Alimentari. L'etichettatura e la Tracciabilità di Filiera come Strumenti di Tutela delle Produzioni Alimentari. Ind Aliment 56, 581:18–22

OSHA (2016) Hazard classification guidance for manufacturers, importers, and employers. Occu-pational Safety and Health Administration (OSHA), U.S. Department of Labor, Washington, DC. Available https://www.osha.gov/Publications/OSHA3844.pdf. Accessed 09 Jan 2020

Parisi S (2012) Food Packaging and food alterations. The user-oriented approach. Smithers Rapra Technology Ltd., Shawbury

Parisi S (2013) Food Industry and packaging materials. User-oriented guidelines for users. Smithers Rapra Technology Ltd., Shawbury

Parisi S, Barone C, Sharma RK (2016) The RASFF: legal bases, aims and procedures for notifications. In: Chemistry and food safety in the EU. Springer, Heidelberg, Germany. https://doi.org/10.1007/978-3-319-33393-9_1

Pellerito A, Dounz-Weigt R, Micali M (2019a) Food sharing and the regulatory situation in Europe. An introduction. in: food sharing chemical evaluation of durable foods. Springer, Cham. https://doi.org/10.1007/978-3-030-27664-5_1

Pellerito A, Dounz-Weigt R, Micali M (2019b) Food sharing in practice: The German experience in Magdeburg. In: Food sharing chemical evaluation of durable foods. Springer, Cham. https://doi.org/10.1007/978-3-030-27664-5_2

Pellerito A, Dounz-Weigt R, Micali M (2019c) Food sharing and durable foods. The analysis of main chemical parameters. In: Food sharing chemical evaluation of durable foods. Springer, Cham. https://doi.org/10.1007/978-3-030-27664-5_3

Pellerito A, Dounz-Weigt R, Micali M (2019d) Food waste and correlated impact in the food industry. A simulative approach. In: Food sharing chemical evaluation of durable foods. Springer, Cham. https://doi.org/10.1007/978-3-030-27664-5_4

Ross MJ, Godin C (2016) Guide entreposage des produits dangereux dans le secteur manufacturier. Multiprévention, Longueil. Available https://multiprevention.org/wp-content/uploads/2018/09/guide-multiprevention-entreposage-produits-dangereux.pdf. Accessed 10 Jan 2020

Roy K (2009) Chemical storage. Sci Teacher 76(7):12

SSTI (2013) Guide pratique d'évaluation et de prévention du risque chimique en enterprise. Fédération régionale des services de santé au travail (SSTI) des Pays de la Loire, Lava. Available https://www.risquechimiquepaysdelaloire.org/sites/default/files/guide_pratique_risque_chimique_ssti_pays_de_la_loire_-_edition_2013.pdf. Accessed 10 Jan 2020

Stellman JM (ed) (1998) The ILO encyclopedia of occupational health and safety, 4th edn. International Labour Office, Genève

Stockwell PB (1996) Automatic chemical analysis. CRC Press, Boca Raton

UN (2019) Globally harmonized system of classification and labelling of chemicals (GHS), 8th revised edn. United Nations, New York and Geneva. Available http://www.unece.org/fileadmin/DAM/trans/danger/publi/ghs/ghs_rev08/ST-SG-AC10-30-Rev8e.pdf. Accessed 09 Jan 2020

Von Rybinski W (2007) Physical aspects of cleaning processes. In: Johansson I, Somasundaran P (eds) Handbook for cleaning/decontamination of surfaces, vol 1. Elsevier B.V, Amsterdam and Oxford, pp 1–55

Ziem GE, Castleman BI (1989) Threshold limit values: historical perspectives and current practice. J Occup Med 31(11):910–918. https://doi.org/10.1097/00043764-198911000-00014

Chapter 4
The Future of Chemicals in the Food Production Ambit

Abstract The continuous and impetuous advance of chemistry in the food and beverage industry is relatively recent and correlated with the creation of new products with enhanced features (e.g., improved durability) on the one side, and the introduction of automated or semi-automated processes similar to non-food industries, meaning also the creation of busy program schedules, the use of 'crowded' areas for manipulation, storage, transportation, etc., and the need of improved machines and equipment which should be functional and cleanable in a relatively low time. Many workers, exposed to chemicals, may suffer critical diseases because of significant adverse effects, even at low doses. The difference in physical state of chemicals (powders, aerosolized dust, gas, etc.) can complicate studies and evaluations. Food industries are not an exception. This chapter explores the future behavior of food production workers when speaking of chemical handling and allowed/notified uses. Probably, food and beverage companies will be forced to gradually consider safety and environmental concerns within their organization, and young/senior managers will be required to comply with safety regulations. Moreover, the increasing number of different (even if similar) food products could be the needed premise for ameliorated logistic procedures in warehouses. In addition, the design of modern foods and beverages could promote the diminution of mixing operations (the 'omnipresent' step in all food and beverage industries) with reduced production steps and risks related to manipulation of chemical products.

Keywords Chemical risk · Cleaning agent · Food additive · Food industry · Safe handling · Sanitizer · Technological aid

Abbreviation

AISS   Association internationale de la sécurité sociale

© The Author(s), under exclusive license to Springer Nature Switzerland AG 2020
R. Chaib and M. Barone, *Chemicals in the Food Industry*,
Chemistry of Foods, https://doi.org/10.1007/978-3-030-42943-0_4

4.1 Chemicals in the Food Production Ambit: Current Perspectives

This book has two main objectives: (i) to show that chemicals are essential for the production of a large number of industrial goods, and food products in particular, and (ii) to promote the culture of work safety.

Chapters 1, 2, and 3 should have displayed the current situation in the ambit of foods and beverages when speaking of the penetration of chemical substances. In brief, the continuous and impetuous advance of chemistry in this field is relatively recent and correlated with:

(1) The creation of new products with enhanced features (e.g., improved durability), with the consequent modification of technical recipes and formulation … by means of the introduction of new ingredients (Pellerito et al. 2019a, b, c, d)
(2) The introduction of automated or semi-automated processes similar to non-food industries, meaning also the creation of busy program schedules, the use of 'crowded' areas for manipulation, storage, transportation, etc., and the need of improved machines and equipment which should be functional and cleanable in a relatively low time.

As a result, chemical substances have gradually invaded the food and beverage environment, but raising concerns about their effects on health are emerged. Many workers, exposed to chemicals, may suffer critical diseases because of significant adverse effects, even at low doses. The difference in physical state of chemicals (powders, aerosolized dust, gas, etc.) can complicate studies and evaluations. Food industries are not an exception.

4.2 Chemicals in the Food Production Ambit: Future Perspectives

On these bases, some prediction may be assumed when speaking of future food and beverage enterprises.

First of all, chemicals will be probably more abundant in the future in relation to food-related activities and correlated procedures (cleaning, sanitization, production of containers and objects for food-contact applications, fuels for thermal machines, etc.). Probably, in a similar situation, and with the increasing complexity of the industrial enterprise, the constant presence of chemicals, and the rapid evolution of little and medium companies into large enterprises, risk assessment will evolve rapidly. Work-related accidents and occupational diseases have a huge impact on workers' health and economic and social benefits. In this situation, companies will be forced to gradually consider these concerns within their organization, and young/senior managers will be required to comply with safety regulations.

On the other side, a possible evolution of safety culture in food industries may concern in future the same nature of products and related food chains. In other terms, it could be anticipated that the increasing separation between different production lines (and products) in the same industry should cause the physical separation between different chemicals—additives, cleaning agents, sanitizers, fuels, etc.—in a logical way (Gurnari 2015; Laganà et al. 2017a, b, c, d, 2019a, b, c, d; Mania et al. 2017). This evolution should be originated because of the continuous differentiation of foods and beverage products in many groups and sub-typologies—organic products, vegan foods, gluten-free edible commodities, regional recipes, etc. Actually, the premise of work safety procedures is already based on factors, such as the separation of chemicals and the safe handling, storage, transportation of these substances and mixtures. However, the increasing number of different (even if similar) food products could be the needed premise for ameliorated logistic procedures in warehouses, and this premise could act as a good tool (Haddad and Parisi 2019a, b).

Finally, the need of 'natural' additives and raw materials to be used in food productions—correlated with the design of modern food products—generally forces producers to search for new substances. Probably, the continuous research will transform the market of chemicals with the resulting creation of intermediate mixtures where a basic raw material (with a traditional process) is coupled with one or more additives with the aim of producing stable ingredients (extended shelf life) with the consequent limitation of operations by food workers at the final production step. In other terms, these productions should limit the number of operating workers when speaking of mixing operations (the 'omnipresent' step in all food and beverage industries) with reduced production steps and possibilities of manipulating intermediates.

References

Gurnari G (2015) Safety protocols in the food industry and emerging concerns. Springer, Cham. https://doi.org/10.1007/978-3-319-16492

Haddad MA, Parisi S (2019a) Evolutive profiles of Mozzarella and vegan cheese during shelf-life. In: 33rd EFFoST international conference "sustainable food systems—performing by connecting". WTC Rotterdam, The Netherlands, 12–14 Nov 2019. Coffee and Poster Session 1, Tuesday, 12 Nov 2019

Haddad MA, Parisi S (2019b) Vegan cheeses VS processed cheeses. Traceability issues and monitoring countermeasures. 33rd EFFoST international conference "sustainable food systems—performing by connecting". WTC Rotterdam, The Netherlands, 12–14 Nov 2019. Coffee and Poster Session 2, Wednesday, 13 Nov 2019

Laganà P, Avventuroso E, Romano G, Gioffré ME, Patanè P, Parisi S, Delia S (2017a) Classification and technological purposes of food additives: the European point of view. In: Chemistry and hygiene of food additives. Springer, Cham

Laganà P, Avventuroso E, Romano G, Gioffré ME, Patanè P, Parisi S, Delia S (2017b) The Codex Alimentarius and the European legislation on food additives. In: Chemistry and hygiene of food additives. Springer, Cham

Laganà P, Avventuroso E, Romano G, Gioffré ME, Patanè P, Parisi S, Delia S (2017c) Food additives and effects on the microbial ecology in yoghurts. In: Chemistry and hygiene of food additives. Springer, Cham

Laganà P, Avventuroso E, Romano G, Gioffré ME, Patanè P, Parisi S, Delia S (2017d) Use and overuse of food additives in edible products: health consequences for consumers. In: Chemistry and hygiene of food additives. Springer, Cham

Laganà P, Campanella C, Patanè P, Cava MA, Parisi S, Gambuzza ME, Delia S, Coniglio MA (2019a) Food gases: classification and allowed uses. In: Chemistry and hygiene of food gases. Springer, Cham

Laganà P, Campanella C, Patanè P, Cava MA, Parisi S, Gambuzza ME, Delia S, Coniglio MA (2019b) Food gases in the European Union: the legislation. In: Chemistry and hygiene of food gases. Springer, Cham

Laganà P, Campanella C, Patanè P, Cava MA, Parisi S, Gambuzza ME, Delia S, Coniglio MA (2019c) Food gases in the industry: chemical and physical features. In: Chemistry and hygiene of food gases. Springer, Cham

Laganà P, Campanella C, Patanè P, Cava MA, Parisi S, Gambuzza ME, Delia S, Coniglio MA (2019d) Safety evaluation and assessment of gases for food applications. In: Chemistry and hygiene of food gases. Springer, Cham

Mania I, Barone C, Pellerito A, Laganà P, Parisi S (2017) Traspa-renza e Valorizzazione delle Produzioni Alimentari. L'etichettatura e la Tracciabilità di Filiera come Strumenti di Tu-tela delle Produzioni Alimentari. Ind. Aliment 56(581):18–22

Pellerito A, Dounz-Weigt R, Micali M (2019a) Food sharing and the regulatory situation in Europe. An introduction. In: Food sharing chemical evaluation of durable foods. Springer, Cham. https://doi.org/10.1007/978-3-030-27664-5_1

Pellerito A, Dounz-Weigt R, Micali M (2019b) Food sharing in practice: the German experience in Magdeburg. In: Food sharing chemical evaluation of durable foods. Springer, Cham. https://doi.org/10.1007/978-3-030-27664-5_2

Pellerito A, Dounz-Weigt R, Micali M (2019c) Food sharing and durable foods. The analysis of main chemical parameters. In: Food sharing chemical evaluation of durable foods. Springer, Cham. https://doi.org/10.1007/978-3-030-27664-5_3

Pellerito A, Dounz-Weigt R, Micali M (2019d) Food waste and correlated impact in the food industry. A simulative approach. In: Food sharing chemical evaluation of durable foods. Springer, Cham. https://doi.org/10.1007/978-3-030-27664-5_4